THE PHYSICAL FOUNDATION OF PROTEIN ARCHITECTURE

THE PHYSICAL FOUNDATION OF PROTEIN ARCHITECTURE

Nobuhiko Saito
Waseda University, Japan

Yukio Kobayashi
Soka University, Japan

World Scientific
Singapore • New Jersey • London • Hong Kong

Published by

World Scientific Publishing Co. Pte. Ltd.

P O Box 128, Farrer Road, Singapore 912805

USA office: Suite 1B, 1060 Main Street, River Edge, NJ 07661

UK office: 57 Shelton Street, Covent Garden, London WC2H 9HE

British Library Cataloguing-in-Publication Data
A catalogue record for this book is available from the British Library.

THE PHYSICAL FOUNDATION OF PROTEIN ARCHITECTURE

ISBN 981-02-4710-9

Printed in Singapore by Uto-Print

Preface

Proteins are biologically important molecules coded on DNA of the genes of living organisms. They play the essential roles for biological functions through their three-dimensional structures. Thus the studies of the structures, which have been undertaken over almost whole range of 20th century by many scientists, are still the central problem of the structural biology. In effect experimental as well as theoretical studies of protein structure and its folding have revealed before us a lot of beautiful and skillful aspects of proteins or Nature herself.

The authors intend to present the folding mechanism of a nascent protein synthesized according to the information of DNA into the complex three-dimensional structure. They also propose an *ab initio* method of prediction of protein structure on the basis of folding mechanism, because they think that these kinds of research provide a fundamental contribution to the structural biology. Now that the human genome project has successfully revealed the whole sequence of the human DNA, and also the DNA sequences of some of other organisms such as *escherichia coli* become available, proteins are the next target of research. The main interest is, besides the experimental determination of three dimensional structures by X-rays, electron microscope or NMR etc, to develop the techniques of rapid determination of protein structures from their sequences, by comparative methods such as homology recognition and fold recognition method, or by theoretical *ab initio* methods.

This book is a revised edition of the review article of the same title appeared in International Journal of Modern Physics B in 1999. In particular a new section on ferredoxin is added, and several rewritings are carried out.

In writing this book the authors tried to begin with rather elementary description to meet the possible requirements from students of physics, chemistry or biology, but assumed the least knowledge of statistical mechanics. The authors hope that this small book could afford to the readers some aspects of wonderful biological world through the window of proteins, and further contribute to open one doorway to the new era of biophysics in 21st century.

This work, in large part, was performed while one of the authors (N.S.) was staying at the laboratory of Professor S.Mitaku of Tokyo University of Agriculture and Technology. N.S. wishes to express his hearty thanks to Professor Mitaku and the members of his laboratory for providing him with facilities and various help. The authors express their gratitude to Professor Y. Oono of the Department of Physics, University of Illinois at Urbana-Champaign for giving the authors the opportunity to prepare this article and various scientific as well as technical advices for improving the manuscript. Thanks are also extended to Professor S. Sugai of Soka University and Dr. T. Yao of Genome Science Center /RIKEN for their useful comments on the manuscript.

Nobuhiko Saitô
Yukio Kobayashi

Contents

Preface   v

Chapter 1   Generalities   **1**

1.1   Introduction   1
1.2   Helix-Coil Transition in Polypeptide   6
    1.2.1 α- and β-keratins   6
    1.2.2   Helix-coil transition   8
    1.2.3   Theory of the formation of α-helices and β-strands
        in protein   10
    1.2.4   Prediction of α-helices and β-strands in a protein   14
    1.2.5   Some preliminary results for the secondary structure
        prediction   15
    1.2.6   Stability of α-helix   18
1.3   Some Aspects of Protein Folding   20
    1.3.1   Reversible denaturation and renaturation: Anfinsen's
        dogma   20
    1.3.2   Hydrophobic core   24
    1.3.3   Folding pathway   27
    1.3.4   Molten globule state   29
    1.3.5   Graphical representations of protein structures   34
    1.3.6   Statistical mechanical simulation of protein
        conformation   36
    1.3.7   Driving force for packing the secondary structures   43

Chapter 2   Mechanism of Protein Folding   **47**

2.1   Island Model   47

2.2 α-Helical Proteins   48

    2.2.1   Sperm whale myoglobin   48

    2.2.2   Cytochrome b_{562}   56

    2.2.3   Sea hare myoglobin   58

2.3   Lysozyme and Phospholipase   62

    2.3.1   Disulfide bonding and lampshade   63

    2.3.2   Lysozyme   65

    2.3.3   Phospholipase   71

2.4   Bovine Pancreatic Trypsin Inhibitor   75

2.5   Flavodoxin and Thioredoxin   84

2.6   Ferredoxin   89

Chapter 3   Folding of a Protein of Unknown Structure   **95**

3.1   *Ab Initio* Method of Prediction of Protein Structure   95

    3.1.1   General principles   95

    3.1.2   Packing order and Anfinsen's dogma   98

    3.1.3   Application to parathyroid-hormone-related protein
(residues 1-34), abbreviated as PTHrP(1–34)   100

3.2   Search for the Conformation of Minimum Energy   104

    3.2.1   Validity of Anfinsen's dogma   104

    3.2.2   Simulated annealing   104

Chapter 4   Topics Related to Protein Structures   **107**

4.1   Phase Transition   107

4.2   Module   108

    4.2.1   Exons, introns, and modules   108

    4.2.2   Stability of modules   108

4.3   Molecular Chaperones   109

4.4   Membrane Proteins   111

4.5   Structure Prediction Based on Protein Data   112

    4.5.1   Secondary structure   113

    4.5.2   Tertiary structure   114

4.6   Concluding Remarks   114

Appendix A   Helix-Coil Transition in Homopolypeptide   117

A.1   Lifson-Roig Theory 117
A.2   Interaction of Side Chains 118
A.3   Conformation in the Helix-Coil Transition Region 119

Appendix B   Levinthal Paradox   121

B.1   Phase Space of a Protein 121
B.2   Ideal Gas 122

Appendix C   Method of Bremermann for Searching the Conformation of Minimum Energy   125

C.1   Outline of the Method 125
C.2   An Example 128

Appendix D   Formation of β-Sheets   131

D.1   Antiparallel β-Structure 131

References   135

Index   143

Chapter 1

Generalities

1.1 Introduction

Living organisms are performing various kinds of biological functions at every stage of their lives. For example, DNA replication, protein synthesis and its regulation, growth, development, differentiation, respiration, digestion, metabolism, material transport, vision, movement, and so on. It is impossible to enumerate them one by one. We are well aware that various proteins are involved in these processes and play important roles. In other words, specific biological functions are ultimately attributed to the specific functions of proteins, which are undertaken by the three dimensional structures of proteins.

Proteins of biological origin, when hydrolyzed, are decomposed into L-amino acids of 20 kinds (Table 1.1) except for some antibiotic polypeptides containing D-amino acid residues, which are considered to be modified from L-amino acids by some reasons. Thus a protein is a linear polymer of L-amino acid residues connected by peptide bonds (Fig. 1.1). The sequence of amino-acid residues is called the primary structure of the protein,[a] which is given genetically by the DNA-base sequence.

It would be be appropriate to give here a brief explanation of DNA.[1] DNA has a double helical structure composed of two polynucleotide strands, one of which has a genetic information described by the sequence of four bases, ade-

---

[a]Hierarchy of protein structure. Primary structure: amino acid sequence numbered from the NH_2-terminus. Secondary structure: α-helix, β-strand, and β-sheet (antiparallel or parallel β-structure) (see Sec. 1.2 and Appendix D). Tertiary structure: three-dimensional structure of a polypeptide chain. Quaternary structure: structure of a protein composed of several subunits having tertiary structures (hemoglobin is an example).

Table 1.1.   Amino acid residues. Only the side groups attached to α-carbons are indicated, excluding proline with internal bond.

Hydrophobic, non-polar

(1) glycine(Gly, G)   (2) alanine(Ala, A)   (3) valine(Val, V)

$$H-\!\!-\!\!-$$

$$H_3C-\!\!-\!\!-$$

$$\begin{array}{c} H_3C \\ \diagdown \\ CH-\!\!-\!\!- \\ \diagup \\ H_3C \end{array}$$

(4) leucine(Leu, L)   (5) isoleucine(Ile, I)   (6) proline(Pro. P)

$$\begin{array}{c} H_3C \\ \diagdown \\ CH-CH_2-\!\!-\!\!- \\ \diagup \\ H_3C \end{array}$$

$$CH_3-CH_2-\underset{\underset{CH_3}{|}}{CH}-$$

(7) methionine(Met, M)   (8) phenylalnine(Phe, F)

$$H_3C-S-CH_2-CH_2-\!\!-\!\!-$$

(9) tryptophane(Trp, W)

Hydrophilic and dissociative

(10) serine(Ser, S)   (11) threonine(Thr, T)   (12) cysteine(Cys, C)

$$HO-CH_2-\!\!-\!\!-$$

$$H_3C-\underset{\underset{H}{|}}{\overset{\overset{OH}{|}}{C}}-\!\!-\!\!-$$

$$HS-CH_2-\!\!-\!\!-$$

(13) asparagine(Asn, N)   (14) glutamine(Gln, Q)   (15) tyrosine(Tyr, Y)

(16) aspartic acid(Asp. D)   (17) glutamic acid(Glu, E)   (18) hystidine(His, H)

(19) lysine(Lys, K)   (20) arginine(Arg, R)

$$\overset{+}{H_3}N-CH_2-CH_2-CH_2-CH_2-\!\!-\!\!-$$

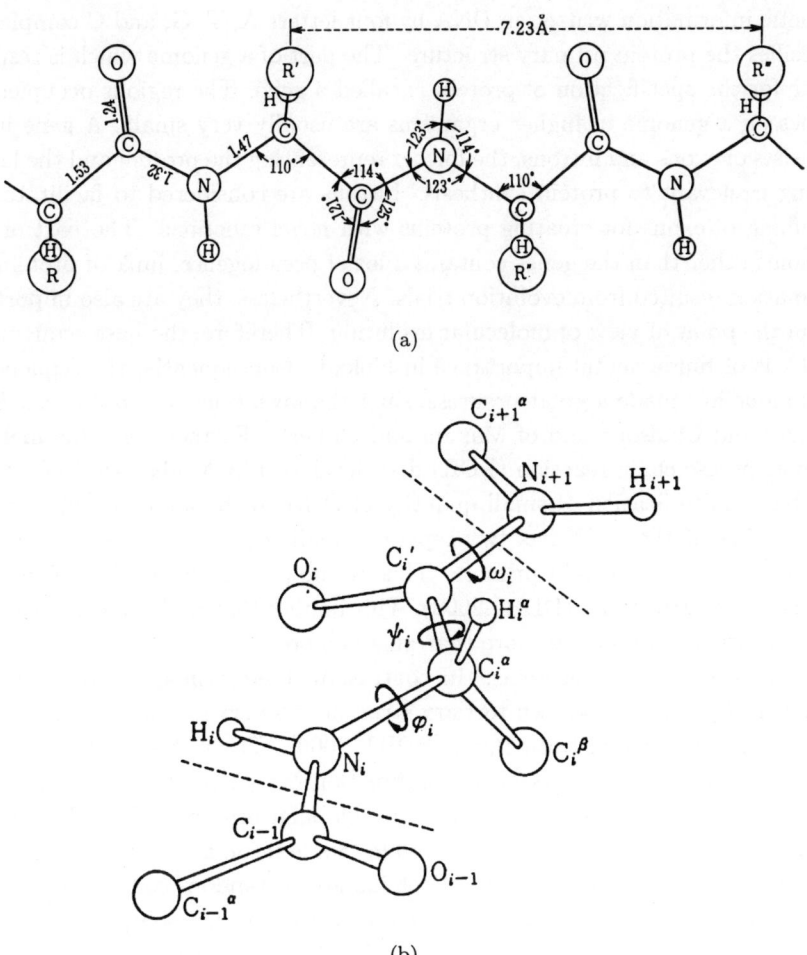

Fig. 1.1.   (a) Polypeptide chain laid on a plane. (b) Peptide bond.

nine (A), thymine (T), guanine (G), and cytocine (C). The molecular genetics has been aiming at elucidating the secret of genetic information, above all, at finding the connection between genetic information and proteins specified by it. This is called *the first deciphering of genetic information* and has been accomplished in the 1960's. We now know that a triplet of bases corresponds to an amino acid, an initiation or a termination of protein synthesis. That is, the genetic information written in DNA by four letters A, T, G, and C completely specifies the protein primary structure. The part of a genome which is responsible for the specification of protein is called a gene. The regions occupied by genes on a genome in higher organisms are usually very small. A gene itself consists of exons and introns, the former representing the protein and the latter being irrelevant to protein synthesis. Introns are considered to facilitate the shuffling of exons for creating proteins with novel functions. The part of the genome other than the genes contains a lot of pseudogenes, junk of protein information resulted from evolution trials. Nevertheless, they are also important from the point of view of molecular evolution. Therefore, the base sequence of DNA is of fundamental importance in biology. Consequently, the sequencing technique has made a great progress, since the inventions of rapid methods of Sanger and Coulson[2] and of Maxam and Gilbert.[3] Furthermore, the method of polymerase chain reaction (PCR) first developed by Mullis (see Ref. 1) has enabled multiplication of small quantity of DNA to the amount sufficient for sequencing, if the DNA has two separate small regions with base sequences complementary to certain primers. Thus we now have almost 7×10^6 entries in the DNA Data Bank (DDBJ,2000). This implies that we have proteins with known primary structures more than this number.

Now we return to proteins. In contrast to their primary structures, the number of proteins of known tertiary structures is very small: only 1.34×10^4 entered in the Protein Data Bank (PDBJ, 2000). The tertiary structure of a protein which carries on its biological function has a complex arrangement of atoms. Proteins soluble in water are called globular proteins, while fibrous proteins such as keratin are usually insoluble. Some insoluble proteins constitute membranes together with lipids. Proteins are in random coil state immediately after synthesized, or at higher temperatures, or in a solution containing special chemical substances, but they can be folded into definite structures under physiological conditions. Proteins can usually be crystallized more easily than artificially synthesized polymers, since their chemical structures specified by the genetic information are completely identical in a sample. Nevertheless, the preparation of crystals suitable for X-ray investigation still requires special techniques. Experimental determinations of the three-dimensional structures of crystallized proteins were achieved first for sperm whale myoglobin and then for hemoglobin in the early 1960's by Kendrew, Perutz, and others, after a long, nearly-20-year, effort of X-ray investigation. In these studies they faced with the difficulties of structure analysis. One of them is the similarity of diffraction powers of C, N, and O atoms, the main constituents of protein molecules. To overcome this difficulty they devised a method to introduce a heavy Hg atom of large diffraction power into a protein molecule. The use of heavy atoms has become a routine procedure in protein X-ray crystallography. Another technical difficulty inherent to this is to prepare crystals of proteins, as mentioned above. Usually X-ray determination of a protein structure is said to require 3 to 4 years of research. Now the synchrotron radiation from a high energy electron storage ring provides a strong X-ray which enables shortening of the experimental time as well as the development of new techniques of X-ray crystallography. Recently, NMR method for structure determination which is done in solution without preparing crystals is being applied, and the full development applicable to large proteins is expected.

The function of a protein is usually performed in aqueous solution. The structure in solutions is supposed to be almost the same as that in its crystalline state. In this connection the NMR study of lysozyme in solution by Smith *et al.*,[4] and the optical activity measurements of crystalline and aqueous lysozymes by Kobayashi *et al.*[5] are of interest. NMR measurements assure that the main chain and many of the internal side chains in solution have the same conformation found in the crystal, but the side chains on the surface are highly disordered among the set of solution structures. The similar conclusion is obtained from the measurement of optical activity, which shows that its magnitude is 19% bigger in crystal than in solution, implying that the constraints of the motion of sides chains in crystal is increased (See Sec. 4.6.(2)).

As mentioned already, the function of a protein is closely related to its tertiary or quaternary structure. Elucidation of the mechanism of protein folding is of a fundamental scientific significance in biology. The determination of the structure from its amino acid sequence is called *the second deciphering of genetic information*. The prediction of the tertiary structure is an urgent technical requirement, resulting in a long waiting list of structure determination, as mentioned above. But currently available methods and facilities are not powerful and numerous enough to meet this requirement. Furthermore now that the human genome project has successfully achieved the determination of all the sequences of human genomes (June, 2000), the next target turns apparently to proteins. The purpose of the present book is, among many attempts to predict the tertiary structure proteins (see Sec. 4.5), first to explain the mechanism of protein folding and to propose a reasonable and possible *ab initio* method of prediction based on the folding mechanism, which we have been studying for these almost two decades. We presented the initial phase of our research at the US-Japan Seminar on self organization of proteins held at Cornell University in 1981. Since then several papers and reports were published in various publications.[6–20] This book is thus intended to provide the fundamentals of protein folding by presenting mainly a systematic explanation of the various researches done by Saitô and his collaborators.

1.2 Helix-Coil Transition in Polypeptide

1.2.1 *α- and β-keratins*

In 1951 Pauling, Corey *et al.*[21] proposed new models of α- and β-keratin which are called α-helix and β-structure. Their proposed structures are shown in Figs. 1.2 and 1.3, that have turned out to be often found as building blocks of a protein, and are called secondary structures. Usually the right(left)-handed α-helix is observed for the poly-L(D) amino acid. The β-structure is a pleated

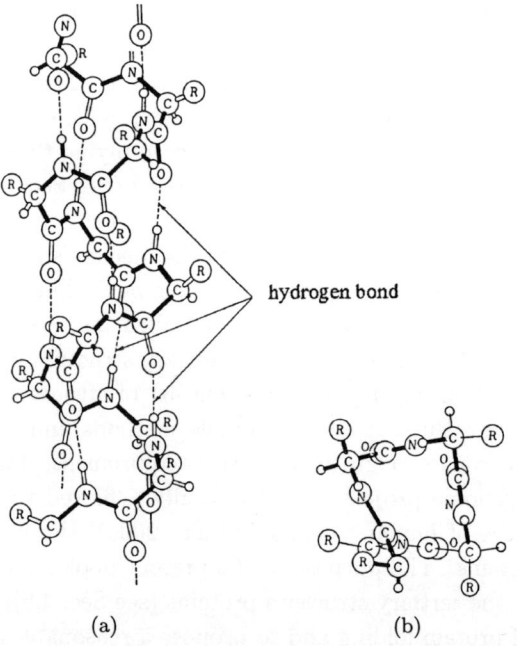

hydrogen bond

(a)　　　　　　　　　　(b)

Fig. 1.2.　Molecular models of α-helix.

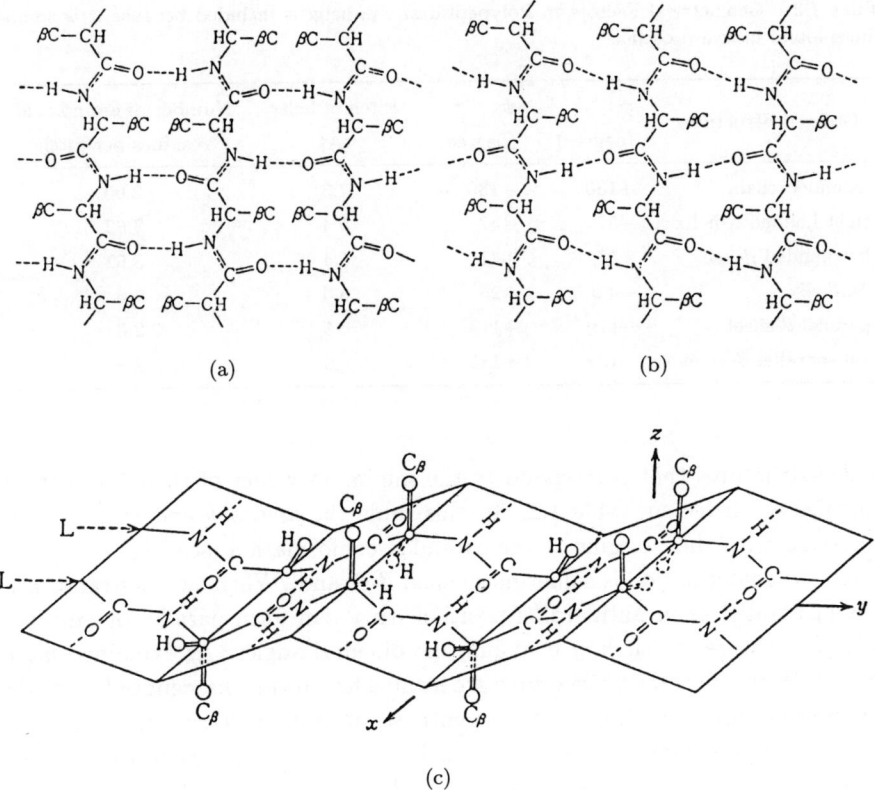

Fig. 1.3. Molecular model β-structures. (a), (c) Antiparallel β-structure. (b) Parallel β-structure.

sheet composed of β-strands in parallel or antiparallel arrangements. The guiding principles for constructing the model of α-keratin are as follows. (i) The peptide bond is planar, or in other words the angle ω is kept at 180°. The molecular parameters such as bond lengths and bond angles are the same as found in small molecules. (ii) The structure obtained by repeated arrangement of similar units is a helix and is stabilized by hydrogen bonds. (iii) The stable structure is searched for among the models satisfying (i) and (ii).

A polypeptide chain can take various conformations such as random coil, α-helix, β-structure, β-strand, etc. These conformations are characterized by the values of their dihedral angles ϕ and ψ. In particular, the values for α-helices

Table 1.2.   Geometrical Factors in Polypeptides. 3_{10}-helix is included because it is sometimes found in real proteins.

Ordered Structure	ϕ (degree)	ψ (degree)	Pitch of helix (Å)	Number of amino acid residues per pitch
extended chain	+180	+180	7.3	2.00
right-handed α-helix	−57	−47	5.4	3.62
left-handed β-helix	+57	+47	5.4	3.62
3_{10}-helix	−4.9	−25	6.0	3.0
parallel β-sheet	−119	+113	6.5	2.0
anti-parallel β-sheet	−139	+135	7.0	2.0

and β-structures that correspond to the minimum values of their interaction energies are listed in Table 1.2. In this table the angles ϕ and ψ are taken clockwise with their origins at the extended conformation (see Fig. 1.1). The extended conformation is sometimes called β-strand. When two β-strands are put in parallel or in antiparallel position, they can form parallel or antiparallel β-structure, by slightly changing the dihedral angles. As explained later (Sec. 1.3), the secondary structures are formed first from a nascent polypeptide chain in random coil state while biosynthesis of protein. Their structures are supposed to be of standard form listed in Table 1.2. In the native proteins, however, the secondary structures are usually deformed from the standard forms. Kabsch and Sander[22] thus presented DSSP (Define Secondary Structure of Proteins) program based on a set of simple and physically motivated criteria for secondary structure through a pattern-recognition process of hydrogen bonded and geometrical features extracted from X-ray coordinates. The DSSP program is available on the Web(http://www.ddbj.nig.ac.jp/). But it cannot necessarily reproduce the standard secondary structures described above (see Sec. 2.6).

1.2.2   *Helix-coil transition*

The stereoregular polypeptides composed exclusively of L-amino acids, such as polyglutamic acid (PGA) or polybenzyl-L-glutamate (PBLG), can be regarded as simple protein models. The solution properties such as viscosity $[\eta]$, ionization I and optical rotation $[\alpha]$, etc., of PGA and PBLG were investigated

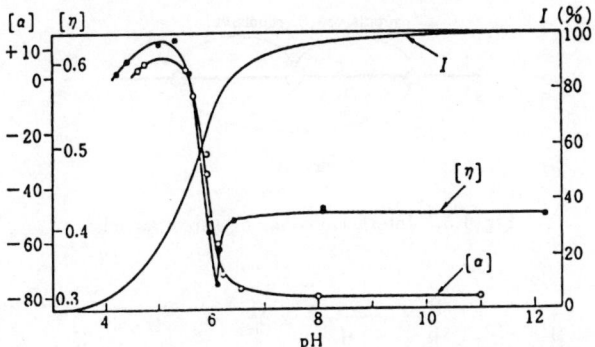

Fig. 1.4.   Helix-coil transition in PGA (reproduced from Ref. 21 with permission).

by Doty and his collaborators.[23] They found that these molecules can take helical structures, identified as α-helices, under appropriate conditions. They further showed that the transition between helix and coil can be induced by changing the solvent conditions, as shown in Fig. 1.4 for PGA. In other words, α-helix first found in solid crystalline state is also stable in solution as a single molecule. We have employed the word transition, because one state changes into another reversibly under quite small changes of solvent conditions, similar to the phase transition between liquid and gas. However, in one-dimensional systems with finite range interaction no phase transition can be expected according to statistical mechanics.[24] The term transition used in statistical mechanics implies a change of state in a mathematically singular manner in the thermodynamic limit, i.e., in the limit of infinitely large number of particles and the volume with the density kept constant. In this sense, although the helix-coil transition in polypeptide is a sharp change of states, it is not, strictly speaking, a phase transition, because it does not have any mathematical singularity. Then, why does transition-like phenomenon occurs in polypeptide? Now consider a one-dimensional lattice gas with N-particles in M sites, where it is assumed that the nearest-neighbor interaction is repulsive and the second-neighbor interaction is zero, but it becomes sufficiently attractive to overcome the nearest-neighbor repulsive forces (Fig. 1.5), when a third particle comes in between. In this system when M (volume) is decreased, some particles have to make contact to form a doublet, resulting in an increase of the pressure, but upon further decrease of M a triplet of particles happen to be made without big increase of pressure. Consequently, the pressure increases steeply only in

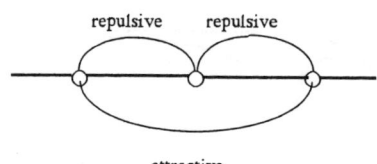

Fig. 1.5.   Interactions among three particles.

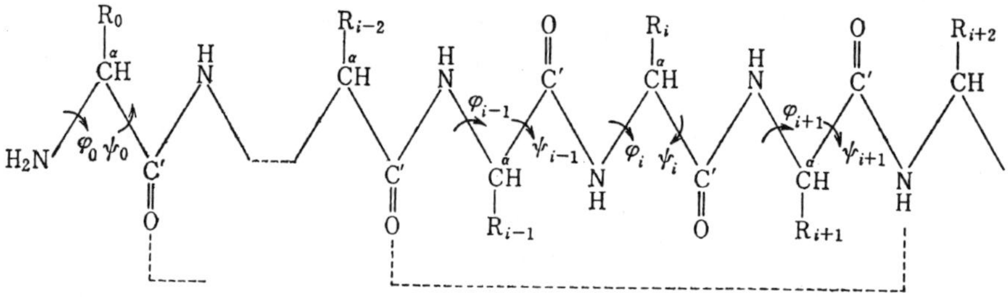

Fig. 1.6.   Polypeptide chain.

the small range of M when it is decreased. This model is a lattice gas version of the helix-coil transition,[25] where forming an initial helix turn is unfavorable, since this conformation occupies a quite small region in the phase space. Once one helix turn is formed, further winding of a helix turn is easy thanks to the hydrogen bonding. In an α-helix of polypeptide a hydrogen bond is formed between the ith and the $(i + 4)$th residues as can be seen from the model of α-helix shown in Fig. 1.6. Theories of helix-coil transition for homopolypeptide were developed by many authors (see Poland and Scheraga[26]) especially by Hill,[27] Zimm and Bragg,[28] and Lifson and Roig.[29] In the next subsection we will present a theory for proteins regarded as heteropolypeptides, in order to apply it for the prediction of secondary structure. The theory reduces to the conventional helix-coil transition theory when applied to homopolypeptides (See Appendix A).

1.2.3   *Theory of the formation of α-helices and β-strands in protein*

For the formation of α-helices, β-strands, and further complex structures in proteins, we propose the island model. An island is defined as a part of the chain which has a definite structure through local interactions among residues. For the present purpose we have only to consider α-helices, β-strands, and coil parts. It is assumed that there exists at least one residue in the coil state between an α-helix and a β-strand and there is no interaction between different islands. In the island model the formation of a structure proceeds first by the birth of embryos or kernels of islands and then by their growth through incorporating neighboring residues successively. The conformation of a polypeptide is described in terms of dihedral angles of two bonds from α-carbons. If a pair of dihedral angles around the bonds from the α-carbon C_i^α are specified, together with the internal rotations of the $(i-1)$th, ith and $(i+1)$th residues, and the structure of planar peptide bonds, they can determine the relative position between O=C_{i-2} of the $(i-2)$th residue and HNC_{i+2} of the $(i+2)$th residue as shown in Fig. 1.6 by $---$. If the relative position becomes appropriate, a hydrogen bond for helix can be formed (see Fig. 1.2). In a β-strand the side chain attached to C_i^α of the ith residue comes close to the side chain of the $(i+2)$th residue (see Fig. 1.3). The interaction is effective only when the dihedral angles around C_i^α, C_{i+1}^α, and C_{i+2}^α take the values appropriate for β-strands. The interaction for α helix or β-strand contribute to the statistical weights $w_2(i-2, i+2)$ and $u_2(i, i+2)$, respectively, which are different from pair to pair in the formulation to follow. By taking account of these statistical weights a statistical mechanical theory using a matrix formulation was developed by Wako *et al.*,[30] but here we present a formalism in the form of recurrence relations which is suitable for computer calculation.[10]

The partition function Z_n of a part of the protein from the 1st to the nth residue is written as:

$$Z_n = Z_n^{(\alpha)} + Z_n^{(\beta)} + Z_n^{(c)}, \tag{1.1}$$

where $Z_n^{(\alpha)}$, $Z_n^{(\beta)}$, and $Z_n^{(c)}$ are the partition functions with the nth residue in α-helix, β-strand, and coil, respectively. We have the recurrence relation:

$$Z_n^{(\alpha)} = \sum_{k=1}^{n} Z_{n-k}^{(c)} H(n-k+1, n), \quad Z_0^{(c)} = 1, \quad Z_0^{(\alpha)} = 0, \tag{1.2}$$

where $H(n - k + 1, n)$ implies the partition function of complete α-states from the $(n - k + 1)$th to the nth residues, written as

$$
\begin{aligned}
H(n &- k + 1, n) \\
&= w_0(n - k + 1)w_1(n - k + 1, n - k + 2) \cdots w_{k-1}(n - k + 1, n) \\
&\times w_0(n - k + 2)w_1(n - k + 2, n - k + 3) \cdots w_{k-2}(n - k + 2, n) \\
&\times \cdots w_0(n - 1)w_1(n - 1, n)
\end{aligned}
\tag{1.3}
$$

in terms of weight functions, $w_0(i)$ for single ith residue and $w_k(i, i + k)$ for a pair of the ith and the $(i + k)$th residues. Similarly,

$$
Z_n^{(\beta)} = \sum_{k=1}^{n} Z_{n-k}^{(c)} B(n - k + 1, n) \,,
\tag{1.4}
$$

where $B(n - k + 1, n)$ is the partition function for complete β-states and can be written as Eq. (1.3) in terms of u for weight functions of β-states. Finally, we have

$$
Z_n^{(c)} = \sum_{k=1}^{n-1} [Z_{n-k}^{(\alpha)} + Z_{n-k}^{(\beta)}] + 1 \,,
\tag{1.5}
$$

and

$$
Z_0^{(c)} = 1, \quad Z_1^{(c)} = 1, \quad Z_1^{(\alpha)} = w_0(1), \quad Z_1^{(\beta)} = u_0(1) \,.
\tag{1.6}
$$

Furthermore, we assume that no interaction exists between the kth and the k'th residues in the same α or β part for $|k - k'| > p$ in α and $|k - k'| > p'$ in β. Thus,

$$
w_k(n, n + k) = 1, \qquad k > p \,,
\tag{1.7}
$$

$$
u_k(n, n + k) = 1, \qquad k > p'
\tag{1.8}
$$

and, consequently,

$$
H(n - k + 1, n) = H(n - k + 1, n - 1)W(n - p, n), \qquad k > p,
\tag{1.9}
$$

$$
B(n - k + 1, n) = B(n - k + 1, n - 1)U(n - p', n), \qquad k > p',
\tag{1.10}
$$

where

$$W(n - p, n) = w_p(n - p, n)w_{p-1}(n - p + 1, n) \cdots w_1(n - 1, n)w_0(n), \quad (1.11)$$

and a similar relation exists for $U(n - p', n)$.

In the following we shall discuss the α state only, because the results can easily be extended to the β state. Equations (1.1)–(1.6) can be employed successively for calculating $Z_n^{(\alpha)}$ for $n < p$. For $n > p$, we have

$$Z_n^{(\alpha)} = \sum_{k=p}^{n} Z_{n-k}^{(c)} H(n - k + 1, n) + \sum_{k=1}^{p-1} Z_{n-k}^{(c)} H(n - k + 1, n), \quad (1.12)$$

where the first term of rhs is

$$T^{(\alpha)}(n) = \sum_{k=p}^{n} Z_{n-k}^{(c)} H(n - k + 1, n)$$

$$= \sum_{k=p}^{n} Z_{n-k}^{(c)} H(n - k + 1, n - 1) W(n - p, n)$$

$$= [T^{(\alpha)}(n - 1) + Z_{n-p}^{(c)} H(n - p + 1, n - 1)] W(n - p, n). \quad (1.13)$$

In a similar way, the recurrence relations for $Z_n^{(\beta)}$ and $Z_n^{(c)}$ can be formulated. These recurrence relations can be used successively for calculating Z_N.

The partition function can also be written as

$$Z_N = \sum_{k=1}^{i} \sum_{k'=1}^{N-i+1} Z_{i-k}^{(c)} H(i - k + 1, i + k' - 1) Z'_{N-k'+1}^{(c)}$$

$$+ \sum_{k=1}^{i} \sum_{k'=1}^{N-i+1} Z_{i-k}^{(c)} B(i - k + 1, i + k' - 1) Z'_{N-i-k'+1}^{(c)}$$

$$+ Z_i^{(c)} Z'_{N-i}^{(c)}, \quad (1.14)$$

where $Z'_{N-i-k'+1}^{(c)}$ implies the partition function for the chain from the C terminus to the $(N - i - k' + 1)$th residue numbered reversely from the C terminus (which is the $(i + k')$th residue from the N terminus) in the coil state.

The probability that the ith residue is in α is given by

$$p_i^{(\alpha)} = Z_N^{-1} \sum_k \sum_{k'} Z_{i-k}^{(c)} H(i-k+1, i+k'-1) Z'^{(c)}_{N-i-k'+1}. \qquad (1.15)$$

Similarly, we have for β

$$p_i^{(\beta)} = Z_N^{-1} \sum_k \sum_{k'} Z_{i-k}^{(c)} B(i-k+1, i+k'-1) Z'^{(c)}_{N-i-k'+1}. \qquad (1.16)$$

1.2.4 *Prediction of α-helices and β-strands in a protein*

In order to calculate the probabilities $p^{(\alpha)}$ and $p^{(\beta)}$ we have to determine the values of statistical weights, w's and u's. They are to be obtained, in principle, from physicochemical calculations, but now we take another method following Wako et al.[30] to determine them so that they can recover the secondary structures of known protein structures. To do this we introduce the objective function F defined by

$$F = \sum_P \left[\sum_{j \in \alpha} \{(1 - p_j^{(\alpha)})^2 + (p_j^{(\beta)})^2 + (p_j^{(c)})^2\} \right.$$

$$+ \sum_{j \in \beta} \{(p_j^{(\alpha)})^2 + (1 - p_j^{(\beta)})^2 + (p_j^{(c)})^2\}$$

$$\left. + \sum_{j \in c} \{(p_j^{(\alpha)})^2 + (p_j^{(\beta)})^2 + (1 - p_j^{(c)})^2\} \right], \qquad (1.17)$$

where the first sum of this equation is taken over all the referring proteins of number N_P. When $F = 0$ is achieved, the prediction for the referring proteins becomes perfect. For our present purpose it would be sufficient to take $p = 4$ (see Appendix A) and $p' = 2$. Then the number of necessary parameters are $20 + 20 \times 20 \times 4 = 1620$ for α-helix and $20 + 20 \times 20 \times 2 = 820$ for β-strand. If we take $p = p' = 4$, we have to determine 3240 parameters by minimizing the objective function F. In the following section, we will present both the cases to see how the determination of the necessary parameters proceeds.

1.2.5 *Some preliminary results for the secondary structure prediction*

The minimization of the objective function has not yet reached the level as we have wished, but the preliminary results obtained up to now should be of some interest for future developments.

We optimized the weight parameters using 80 proteins (not listed here) which lack sequential homology among them, and then estimated the prediction accuracies for 13 proteins (shown in Table 1.3, column (a)) other than the 80 reference proteins. Firstly, we minimized the objective function F by changing all of 3240 parameters. The estimation results for the 13 proteins indicates that the overall accuracy (= total number of correctly predicted residues (α, β and coil) in 13 proteins/total number of residues of the 13 proteins) is 69.6%, while the accuracies of α-helices (ACR(α)) and β-strands (ACR(β)) (= total number of correctly predicted residues in α-helices (or β-strands) of 13

Table 1.3. Prediction accuracy for 13 proteins for estimation. Column (a), all of 3240 parameters were optimized. Column (b), the values of the parameters for $k = 3$ and 4 in β-strands were set at 1.

PDB-ID	No. of Residues	Accuracy	
		(a)	(b)
1ACX	108	59.259	62.037
1CTE	68	64.706	52.941
1LH1	153	84.314	64.967
1UBQ	76	67.105	77.632
2ALP	198	53.030	51.010
2CDV	107	75.701	77.570
2CI2	65	69.231	58.462
2WRP	104	57.692	56.731
4RHV2	255	69.020	65.882
3CLN	143	67.832	62.937
4FXN	138	91.304	89.855
5LYZ	129	68.217	64.341
7RSA	124	74.194	72.581
Total	1668	69.425	67.626

proteins/number of residues observed in α-helices (or β-strands) of the 13 proteins) are 66.3% and 46.9%, respectively. These numbers are more important than the 69.6%, because the latter includes a big contribution of the correctly predicted coil states.

The reasons for the poor accuracy, especially in β-strands, are as follows:

(1) 457 pairs among 3240 pairs of the 13 proteins are not observed in the secondary structures of the 80 proteins. Thus the parameters for the missing pairs cannot be optimized properly.

(2) The distance of the interactions was taken $k \leq 4$ in case of β-strands as well as in case of α-helices, but the statistical weights of interacting pairs of the ith and the $(i \pm k)$th residues ($k = 3, 4$) may not be significantly different from 1 in β-strands. Thus, the parameters for these pairs may be estimated artificially and erroneously during the optimization process.

The number of correctly predicted residues depends on the degree of the optimization of the parameters corresponding to the pairs necessary for prediction. We hoped that the above mentioned accuracy would be improved by assuming $k \leq 2$ for β-strands and by reducing the number of amino acid pairs in the proteins for optimization. Thus, we optimized again the weight parameters, assuming $k \leq 4$ for α-helices and $k \leq 2$ for β-strands. We assigned unity as the value of the parameters for $k = 3$ and 4 in β-strands and performed the optimization with these parameters fixed. As Table 1.3(b) shows, the prediction accuracies for the 13 proteins fall between 51% and 90%, and the average accuracy of the proteins for estimation is 68.1%, with $\text{ACR}(\alpha) = 65.4\%$, and $\text{ACR}(\beta) = 45.9\%$. These values are lower than those of Table 1.3(a), where all of 3240 parameters are optimized. The prediction accuracy of the 80 proteins for the parameter optimization is only 78.2% in (b), but it is 79.2% in (a).

This suggests two features. First, the parameters, 1620 for α-helices and 820 for β-strands, have not yet been optimized sufficiently. As Table 1.4 shows, the number of the residues predicted correctly in the 80 proteins for the parameter optimization increases as the optimization process progresses. The prediction accuracy of β-strands was remarkably improved. Second, a β-strand usually makes a β-sheet with neighboring β-strands. This fact may introduce some indirect interactions between $k = 3$ and 4 pairs, resulting better prediction accuracy than the case $k \leq 2$. Or, speaking physically, since the structures are sometimes susceptible to slight changes while making a β-sheet

Table 1.4. Improvement of prediction accuracy during the process of the parameter optimization. The values of the parameters for $k = 3$ and 4 in β-strands are set at 1. These results show the numbers of the residues correctly predicted and those incorrectly predicted for the 80 proteins for the parameter optimization. NR is the total number of residues of these 80 proteins. NHH is the number of the residues correctly predicted as α-helix, NHE is the number of the residues incorrectly predicted as β-strand, though they are actually in α-helix and NCH is the number of the residues incorrectly predicted as α-helix, though they are actually in coil. The other notations are analogous to these. ACR is the prediction accuracy. As the parameter optimization proceeded, as indicated from the 2nd to 4th lines, the prediction accuracy, especially of β-strands is found to have improved.

NR	NHH	NEH	NCH	NHE	NEE	NCE	NHC	NEC	NCC	ACR
11778	2035	67	314	30	1771	467	617	957	5520	79.182
11778	2048	91	352	60	1351	549	574	1353	5400	74.707
11778	2053	98	335	58	1527	575	571	1170	5391	76.167
11778	2082	83	347	42	1709	525	558	1003	5429	78.282

(Sec. 1.2.1), folding proceeds with the standard β-strand not necessarily the same as in the crystalline state, but take the native structures after making a β-sheet. Similar features like this are often observed in the actual process of folding (see Secs. 2.2.3 and 2.6). The secondary structures are usually more or less deformed at the final stage of folding. Furthermore, antiparallel β-sheets must be determined through checking whether the neighboring two β-strands have the ability to make an antiparallel β-sheet (see Appendix D). Again they are usually deformed. Thus we reach the conclusion that the numerical values of the statistical weights, w's and u's are for standard structures, and for this purpose one has to prepare referring proteins having standard secondary structures. Practically it is difficult, if not impossible. This may be a possible reason for the unsatisfactory prediction accuracy of the secondary structures, especially of β-strands. This consideration holds for empirical methods of prediction to be discussed in Sec. 4.5.1.

The precision of order ca.70% in Table 1.3 seems to be the upper limit of any method for secondary structure.

We should note that various methods of the secondary structure prediction have been developed mostly by the homologies between the primary structures and by the statistical inference based on protein data base, and their prediction accuracies have reached about 70%.[112,113] That is, our method gives results comparable to them (see Sec. 4.5.1). In this connection, we refer to the proposal of Rost, Sander, and Schneider.[161] They showed that considerable variation in the position and length of secondary structure segments can be accommodated within the same three-dimensional structure. Thus, the goal of the prediction accuracy, they say, can be reduced to some extent, and a new measure of segment overlap is introduced to compromise between permissiveness and precision. But it is unknown whether or not the accommodated secondary structures are consistent with the folding mechanism to be discussed in Sec. 1.3. We think that more effort is required to improve the accuracy of secondary structure prediction.

1.2.6 *Stability of α-helix*

The conformational energy of an α-helix is composed of several energies; for example, rotational energies around the dihedral angles, nonbonded interactions between the atoms in the chain, electrostatic energies and hydrogen bonds. Among them the hydrogen bonds between two amino acid residues that are 3 residues apart (in other words at a medium distance) in the main chain play the decisive role for the formation of the helix as can be seen in the theories of helix-coil transition. The contribution of various energies mentioned above in the conformation of poly-L-alanine in the crystalline state was calculated by Kosuge *et al.*[31] improving the result of Ooi *et al.*[32] Table 1.5 gives their results and shows that the largest contribution comes from the nonbonded interactions among mostly nearest neighbors which cannot, by themselves, hold stably the helical structure. On the other hand, the contribution from the hydrogen bonds is smaller but is most important for the stability as explained below. In the case of helix formation the hydrogen bond is directional, short-range and

Table 1.5. Conformational energy per residue of the α-helical structure of poly(L-alanine).

	Energy (kcal/mol-res)	
	Ooi *et al.*	Kosuge *et al.*
Rotational	0.49	0.58
Nonbonded	−5.99	−7.15
Electrostatic	−1.10	2.60
Hydrogen bond	−1.74	−1.02
Total	−8.34	−5.00

medium-distance interaction in our terminology to be defined in Sec. 1.3.3. It is noted that the hydrogen bond energy can easily disappear, while other interactions do not change significantly, when the helical structure is slightly deformed, as shown in Fig. 1.7.[31]

This fact implies that the largest nonbonded interactions, if alone, cannot hold the helical structure. The energy function of a hydrogen bond has a

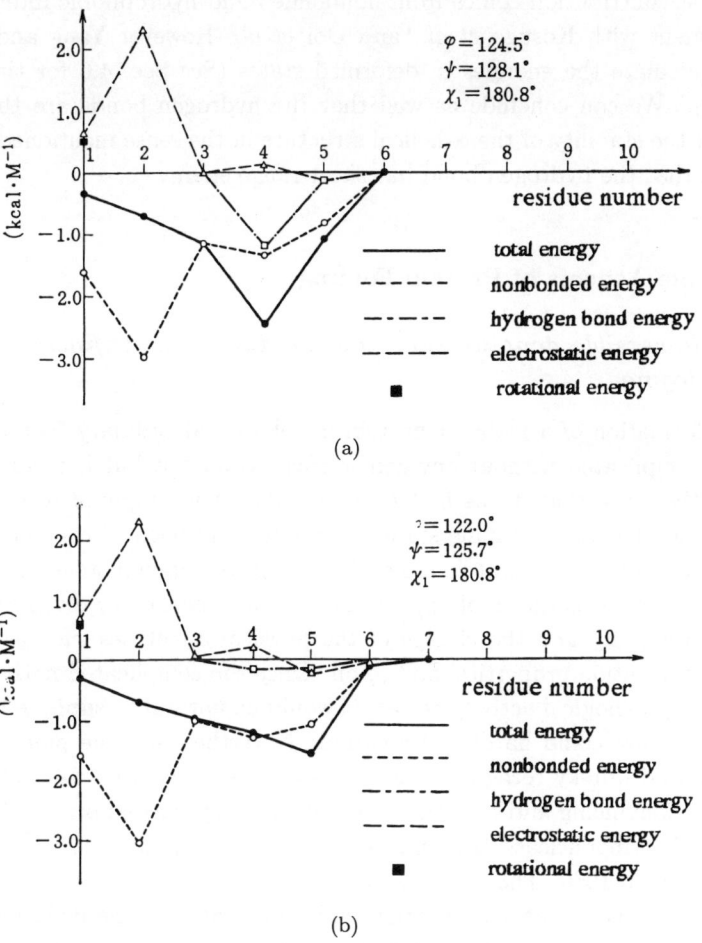

(a)

(b)

Fig. 1.7.   Distribution of energies among residues (reproduced from Ref. 31 with permission). (a) α-helix. (b) Slightly deformed α-helix.

sharp minimum with respect to the atomic distance and the bond is thus easily broken by a slight deformation of the helical structure. Even if their energy is smaller than the nonbonded energy, hydrogen bonds are essential for keeping the α-helix, and bring about the stability. Although the above calculations are performed in vacuum, recent elaborate calculations of interaction energies in α-helix by taking account of solvent effect by Yang and Honig[33] show that the numerical values are not qualitatively different from those in vacuum and yet the largest contribution comes from nonbonded and hydrophobic interactions in agreement with Kosuge *et al.* and Ooi *et al.* However Yang and Honig do not calculate the energies in deformed states (See Sec. 4.6 for the effect of water). We can conclude as well that the hydrogen bonds are the main factor for the stability of the α-helical structure in the sense mentioned above, provided that the hydrogen bond has short-range energy.

1.3   Some Aspects of Protein Folding

1.3.1   *Reversible denaturation and renaturation: Anfinsen's dogma*

The conformation of a globular protein in solution at ordinary temperatures is quite complicated without any geometrical symmetry, but it is an ordered state in the sense that it has biological activity. It is supposed to be almost the same as that of crystalline state as already mentioned. This complicated conformation of a single protein molecule is destroyed upon increasing the temperature or by the addition of appropriate chemical agents, as revealed by the loss of its activity and the change of the physical quantities such as optical properties, solution properties, and so on. Once the complicated native structures having biological activity is lost, it would be natural to suppose that the native structure could hardly be restored. Nevertheless, some pioneers such as Anson and Mirsky recognized as early as in 1925 that this was not always the case. Convincing and beautiful experiments were carried out by Anfinsen *et al.*[34,35] for ribonuclease and, independently, by Isemura *et al.*[36,37] for taka-amylase around 1960. Their surprising experimental facts demonstrate clearly the reversible nature of denaturation and renaturation. The denatured proteins can recover the biological activities and their complicated conformations, when their respective conditions of the solution are restored. Isemura, Takagi and others,[37,38] furthermore, were able to obtain a crystalline state from the

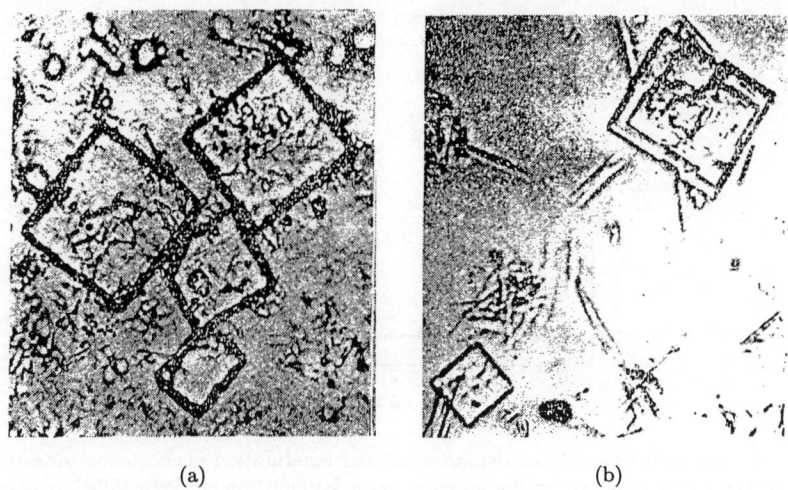

(a)   (b)

Fig. 1.8. Recrystallization of denatured taka-amylase (reproduced from Ref. 30 with permission). (a) Crystals of native taka-amylase. (b) Crystals obtained from denatured and inactive taka-amylase.

solution of denatured taka-amylase, which, when resolved again in solution, can exhibits the activity (Fig. 1.8). The reversible phenomenon thus discovered is quite important in the physicochemical studies of protein. It implies that the change of conformations of a protein is governed by the law of thermodynamics, and especially that the native conformation of a protein is the state of least free energy in the biologically significant circumstances. This statement is *the first principle of protein folding* from equilibrium theoretical point of view, and is called Anfinsen's dogma. Another aspect of reversible renaturation phenomena is observed when some degenerating agent, such as urea or guanidine hydrochrolide is added slowly or the temperature is increased gradually. The protein does not suffer any denaturation up to a certain point, but beyond this point an abrupt degeneration takes place, as shown in Fig. 1.9.[38] It is a diffuse first order phase transition having a sigmoidal shape, which is considered as an evidence of the existence of a cooperative interaction (see Secs. 4.1. and 1.3.6). Thus sigmoidal shapes are observed in helix-coil transition(see Appendix A). In the case of helix-coil transition, the system is not a mixture of random coil and helix at the transition region (see Appendix A).

In proteins, the conformations are governed by long-distance interactions with non-vanishing potential which can yield first order phase transition in

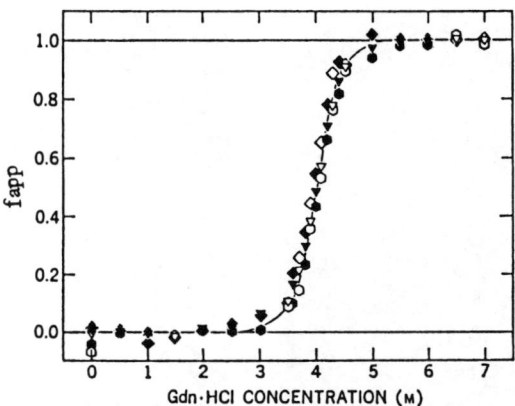

Fig. 1.9. The Gdn-HCl induced denaturation and renaturation phenomenon measured by optical rotation of lysozyme in the absence (open symbols) or presence (filled symbols) of Ca^{2+}. The values of f_{app} were calculated by Eq. (3.4) (see below) from ellipticity at 289 nm ($\diamond$, $\blacklozenge$), 255 nm ($\circ$, $\bullet$) and 222 nm ($\triangledown$, $\blacktriangledown$). All the data lie on a single curve. This is a peculiar case of lysozyme. The curves of f_{app} for different wave lengths do not always coincide with each other as in the case of α-lactalubumin (see, Fig. 1.13) (reproduced from Ref. 39 with permission).

the limit of infinitely large molecules, as shown in the statistical mechanical theory of protein conformation[82] (see Sec. 1.3.6). The transition between the native (N) and the degenerate state (D) (N-D transition) is thus described approximately by[39]

$$N \rightleftharpoons D \qquad\qquad (1.18)$$

and the transition temperature T_m is given by the condition $\Delta G = 0$ or $T_m = \Delta H/\Delta S$, where ΔG, ΔH and ΔS are differences of the Gibbs free energy, heat content and entropy between D and N respectively. Since T_m must be positive, both ΔH and ΔS are either positive or negative. In the former case, N state is stable below T_m and in the latter case N state is stable above T_m (inverse transition). This is quite similar to the first order phase transition between ice and water, where no intermediate state is observed. At the transition temperature two states of ice and water coexist but their relative amount cannot be determined without specifying some extensive quantity such as volume or free energy other than the intensive quantities of temperature and pressure. When we consider the transition (1.18) as a two-state equilibrium arising from a (diffuse) first order phase transition, and introduce the equilibrium constant K, which is equal to

$$K = \frac{n_D}{n_N} = \exp\left(-\frac{\Delta G}{RT}\right) \qquad (1.19)$$

where $n_D(n_N)$ is the mole fraction of D (N). ΔG must be zero at the transition point or at the middle of the transition region as mentioned above. Thus two states N and D can coexist. This situation really occurs in some proteins (see Tanford[40]) as shown in Figs. 1.9 and 1.13, differing from the helix-coil transition. This situation can takes place when the intermediate states at the transition region are unstable. If they are stable, however, one observes the molten globule states.

Now let us return to the principle of protein folding of equilibrium theoretic view point. Levinthal[41] doubted on the statistical thermodynamical approach, and proposed a sequential pathway. To understand this we have to reconsider the conformation space. In statistical mechanics, the phase space is composed of the momentum space and the configuration space. We simply assume, as usually done in polymer physics, that these two spaces can be separated and consequently we exclusively discuss the conformation space (the configuration space of a single molecule). The conformation change of a protein is brought about by changing the dihedral angles of all the amino acid residues while the bond lengths, bond angles, and the torsional angles (ω) are kept unchanged because their motions have time scales much faster than those of conformation changes. The rotation of the dihedral angles are performed under the hindrance potentials with three minima, one trans and two gauche positions.

By discretizing the whole conformation space in terms of the three minima of each internal rotation, the number of the conformations available for a protein with $n + 1$ amino acid residues amounts to $3^{2n} = 10^{2n \log 3} \sim 10^{0.9n}$. If one searches for the state of the minimum energy by surveying all the possible conformations, assuming each state requires 10^{-20} s, a time much smaller than the period of molecular vibration, for an $n = 100$ polypeptide, it takes almost 10^{70} s which is much longer than the life of the Universe since the Big Bang occurred 10^8 years $= 10^{17}$ s ago. On the other hand the folding takes place within a very short time. This fact was first pointed out by Levinthal in May 1967 at the symposium on *"Macromoléules hélicoïdales en solution"* in Paris and is called Levinthal paradox. Actually the estimate of the required length of time did not appear in the above Levinthal's abstract[41] but Jaenicke[42] (see also Levinthal[43]) reported the time as 10^{60} times larger than the experimentally observed time. The time for the secondary structures to form is the order of seconds and that of the tertiary structure is of the order of minutes (see Sec. 1.3.3). This is remarkably quick compared with the above estimate of 10^{70}s. Otherwise, nascent proteins in *vivo* could not carry out their biological functions at the proper time. This quickness is *the second principle of protein folding* from the kinetic point of view. From the above considerations we have three rather inconsistent aspects of protein folding, i.e., the first and the second principles from the equilibrium and kinetic points of view and the Levinthal paradox. They can be reconciled, if the proteins fold in their native states not randomly, but along almost definite paths. It implies that the search for the conformation of the minimum energy is carried out in a restricted space rather than in the whole conformation space. The reversible denaturation-renaturation phenomenon established by Anfinsen and Isemura takes place in this restricted region and is supposedly quick in this small restricted space. Consequently, the uniquely folded structure thus obtained is not necessarily the lowest free energy structure as pointed out by Levinthal. This fact will be discussed later (see Secs. 3.1.2 and 3.2.1). The statistical mechanical meaning of Levinthal paradox is discussed in Appendix B. How the restricted conformation space can be found is the main topic of the following Secs. 1.3.2 and 1.3.3.

1.3.2   *Hydrophobic core*

Hydrophilic amino acids with charged groups or polar groups (see Table 1.1) are soluble in water, but nonpolar hydrophobic amino acids are usually insoluble. Therefore the hydrophilic amino acid residues appear on the surface of a protein molecule, while the hydrophobic ones are located inside the molecule, forming the hydrophobic core. This important feature was pointed out in 1959 by Kauzman.[44] Thus, the nature of hydrophobic interactions and their relevant properties have been extensively investigated by many researchers, in particular by Némethy and Scheraga.[45] Water at ordinary temperatures has a disordered ice-like structure, where some of the hydrogen bonds are broken due to the thermal motion, while others still bind water molecules. Around a hydrocarbon group in water, water molecules are under the influence of non-directional van der Waals force instead of highly directional force due to hydrogen bonds. This situation turns out to hinder the thermal motion of the water molecules. This facilitates further formation of hydrogen bonds than in pure water. Consequently, there occurs a region of ice-like structures with small entropy around the hydrocarbon group. When a hydrocarbon molecule or group is brought into water, this process is slightly exothermic, because the van der Waals interaction makes decrease of the internal energy, and overcomes the loss of hydrogen-bond energy, resulting in the decrease of enthalpy. But the free energy itself increases because of the entropy decrease of water due to the formation of ice-like structure. This implies that the hydrocarbons are insoluble or poorly soluble in water and thus hydrophobic. Nozaki and Tanford[46] and later Jones[47] determined the hydrophobicities of all the residues, observing the free energy changes accompanied by the replacement of water to organic solvents. Their results are shown in the second column of Table 1.6. Looking

Table 1.6. Characteristics of amino acid residues. The average r_G, average θ and van der Waals radius defined in Sec. 2.2 are included.

Amino acid residue	Hydrophobicity index (kcal/mole)	r_G (average, Å)	θ (average, deg.)	van der Waals radius (Å)
1. glycine	0.10	1.969		
2. alanine	0.87			
3. valine	1.87	1.969	12.472	2.72
4. leucine	2.17	2.069	28.109	2.98
5. isoleucine	3.15	2.284	17.783	2.98
6. proline	2.77	1.878	41.291	3.16
7. methionine	1.67	3.181	31.240	2.93
8. phenylalanine	2.87	3.426	41.526	3.16
9. tryptophan	3.77	3.859	46.323	3.44
10. serine	0.07	1.954	22.981	1.97
11. threonine	0.07	1.945	14.564	2.44
12. cysteine	1.52	2.395	30.117	2.36
13. asparagine	0.09	2.538	33.889	2.37
14. glutamine	0.00	3.231	28.464	2.73
15. tyrosine	2.67	3.890	45.371	3.19
16. aspatic acid	0.66	2.557	33.549	2.34
17. glutamic acid	0.67	3.232	30.793	2.71
18. histidine	0.87	3.188	41.031	2.84
19. lysine	1.64	3.636	31.201	3.05
20. arginine	0.85	4.319	27.531	3.09

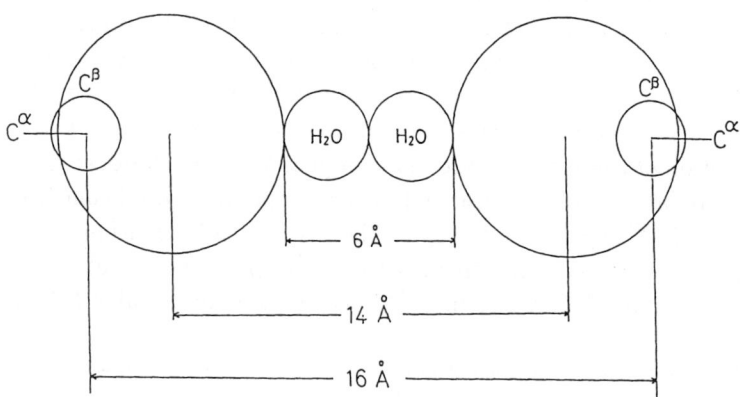

Fig. 1.10. Hydrophobic interaction.

at this table, we classify Trp (W), Ile (I), Phe (F), Leu (L), Val (V), and Met (M) as strong hydrophobic residues and Tyr (Y), Cys (C), and Ala (A) as weak hydrophobic ones. Pro (P) and Lys (K) are excluded, because the latter is basic, and the former is usually located at the surface of a protein molecule, not taking part in the hydrophobic interaction but bending the chain. Tyr and Cys is classified as weak, but may be omitted since they have a polar group OH or SH. When two hydrocarbons come closer, the ice-like structures overlap, and the total ice-like region becomes smaller compared with the case when they are separated sufficiently. Consequently, the hydrophobic interaction is attractive and becomes effective when two ice-like regions begin to overlap. This interaction is fairly long range. If we take 4 Å for the radius of a hydrocarbon group and 3 Å for the thickness of the ice-like layer, then 14 Å is the range of the hydrophobic interaction (see Fig. 1.10). Experimentally, this interaction was shown by Israelachvili and Pashley[48] to be exponentially decaying with a decaying length of 10 Å. Thus, we may express the hydrophobic interaction as follows for the convenience of computer calculation:

$$u(r) = \begin{cases} u_0, & r \leq r_0 \\ u_0 \exp[-(r - r_0)/d], & r \geq r_0 \end{cases} \tag{1.20}$$

where r is the spatial distance in three-dimensional space between the hydrophobic residues and r_0 is the sum of the van der Waals radii of the hydrophobic side chains. We can tentatively assume $u_0 = -3.0$ kcal/mol for pairs of strong hydrophobic residues, -1.5 kcal/mol for pairs of strong and weak hydrophobic residues, and zero, otherwise. The decaying length d may be different from pair to pair but is assumed to be 10 Å irrespective of pairs for simplicity. The hydrophobic interaction enables the formation of the hydrophobic core and plays an essential role in the protein folding as described below. The main point which will be shown there is that the hydrophobic core is made not by a simple aggregation of hydrophobic residues but by their specific combinations.

1.3.3 *Folding pathway*

Now, we can propose the folding pathway, details of which will be described in the pages to follow, but an outline is given here. The first event is the quick formation of the secondary structures in standard forms (α-helices, β-strands, and antiparallel β-sheets, setting aside parallel β-sheets which will be discussed later in Sec. 2.5). This event is considered to be quick, because the secondary structure arises from the interactions between the pair of residues at short distance,[b] i.e., between neighboring amino acid residues on the chain. A statistical mechanical theory of their formation is already discussed in Sec. 1.2.3. The antiparallel β-structures are then formed between the neighboring β-strands. How these are formed, however, is still an unsolved problem. Possible mechanisms will be discussed in Appendix D. The transient states composed almost exclusively of the secondary structures were really found experimentally by several authors and are called the molten globule states which will be discussed in Sec. 1.3.4.

The next process is the packing of the secondary structures thus constructed. The nature of hydrophobic interactions was discussed in detail in Sec. 1.3.2. With this in mind, we can propose that the driving force for the packing of the already formed secondary structures is the long-range

---

[b]In this book the words "distance" and "range" are distinguished when used in connection with interaction. The "distance" is used for the numbers of the amino acid residues intervening along the chain, and the "range" is used for the three-dimensional range of the interaction between the residues. For example, the electrostatic Coulomb force is a long range force, and we sometimes consider the short range interaction of van der Waals type between two atoms at a long distance. When necessary, the phrases "three-dimensional distance" or "spatial distance" is employed between two amino acid residues to avoid the unusual usage of "range" defined above.

hydrophobic interactions between the nearest hydrophobic residues (usually, at medium distance) after the formation of the secondary structures. This is because an efficient and quick packing of the secondary structures must be carried out without making wrong folding repeatedly. The binding between the nearest hydrophobic pairs is quicker without too many trials, in other words, with less loss of conformational entropy of the part between them than between other long-distance pairs. The folding into the correct structure without failing is achieved by long-range interactions between medium-distance pairs, because any short-range interaction such as the Lennard-Jones potential acting only between residues located nearby cannot determine a stable structure, since small changes of conformations are possible without bringing about an appreciable change of interaction energy (see Sec. 1.2.6). The important role of the Lennard-Jones potentials is to prevent the collapse of the molecule and to maintain its volume. Moreover, the interaction energy at the contact distance of two residues by hydrophobic binding is $-3 \sim -4$ kcal/mole, while that of the Lennard-Jones potential is about -0.2 cal/mole. Coulomb forces usually considered to be long range, however, may not be effective for packing because the big dielectric constant of surrounding water reduces the forces. In this connection it is noted that the essentially important point in quick and correct folding into the native structure is the distribution of hydrophobic residues on the chain and not the indivisual hydrophobic residues. In fact Wu and Kim[49] replaced all the hydrophobic residues with leucine in the α domain of α-lactalbumin, with the result that it yielded the molten-globule similar to that of the native one.

We conclude that the long-range interaction indispensable for protein folding is the hydrophobic interaction which can make contacts between nearest medium-distance pairs. The effectiveness of long-range and medium-distance hydrophobic interactions is also clarified in Sec. 1.3.7. The importance of the hydrophobic interaction was already pointed out by many researchers, especially by Kauzman[44] as already mentioned in Sec. 1.3.2 and recently by Srinivasan and Rose[50] and by Sun, Thomas and Dill.[51] However, their long-range nature discussed here and the specificity in choosing the partner of interaction, which will be discussed in Sec. 1.3.7, were unnoticed. They are the key factors in protein folding as will be shown in Chapter 2, and by enabling to form specific hydrophobic core, effectively restrict the conformation space for realizing the second principle of protein folding, avoiding Levinthal paradox.

1.3.4 *Molten globule state*

The two-state description of the protein conformation change as shown in Fig. 1.9 does not necessarily hold in many proteins. Rather intermediate structured are found in a lot of proteins such as α-lactalbumin, carbonic anhydrase, and cytochrome c, etc. The researches were performed, in kinetic as well as in equilibrium ways, by a variety of methods, for example, by circular dichroism (CD), nuclear magnetic resonance (NMR), fluorescence, hydrogen exchange, X-ray, solution properties, and so on. We do not enter into the details of various experimental methods of this field and ask the readers to refer to the review articles such as by Ptitsyn,[52] Privalov,[68] Kuwajima,[53] Kim and Baldwin,[54] Dill and Shortle,[55] Arai and Kuwajima,[56] Sugai and Ikeguchi.[57]

In particular, the CD spectra in the near ultra-violet ($270 - 290$ nm) region come from the aromatic domain, indicating aromatic groups (Trp and Phe) immobilized as in the tertiary structure, while those in the far ultra-violet (222 nm) region from peptide domain indicate the presence of α-helices.

We first discuss some of the kinetic studies. Kuwajima, Sugai and their collaborators[59] examined the refolding processes in apo-α-lactalbumin and lysozyme by CD measurements following upon rapid mixing of the unfolded protein in 6M guanidine hydrochrolide (GdnHCl) with water. The results are shown in Fig. 1.11, where one sees that during the dead time (< 0.5s) of measurements the tertiary structures observable by $[\theta]_{270}$ for lactalbumin and $[\theta]_{290}$ for lysozyme at near ultra violet are not yet been formed, but the formations of the secondary structures observed by the ellipticity $[\theta]_{222}$ are almost completed, with long tails approaching $[\theta]_{\infty}$ at equilibrium. As time goes on, the tertiary structures approach their equilibrium structures gradually. The rapid mixing experiments mentioned above revealed the existence of the kinetic molten globule state having the native secondary structures. Recently much more rapid reaction techniques such as stopped-flow CD have been developed and have enabled to study the rapid refolding of proteins (see for example, Ref. 56). In particular Chaffotte *et al.*[58] were able to shorten the dead time less than 4msec and observed the formation of nativelike secondary structures in the burst phase of refolding in guanidine-unfolded hen egg-white lysozyme.

The kinetic molten globule state implies that the secondary structures are formed first almost completely, and their packing proceeds gradually to form the tertiary structure. This is what we have assumed in the folding path in Sec. 1.3.3, and this situation is also convenient for finding the driving force for

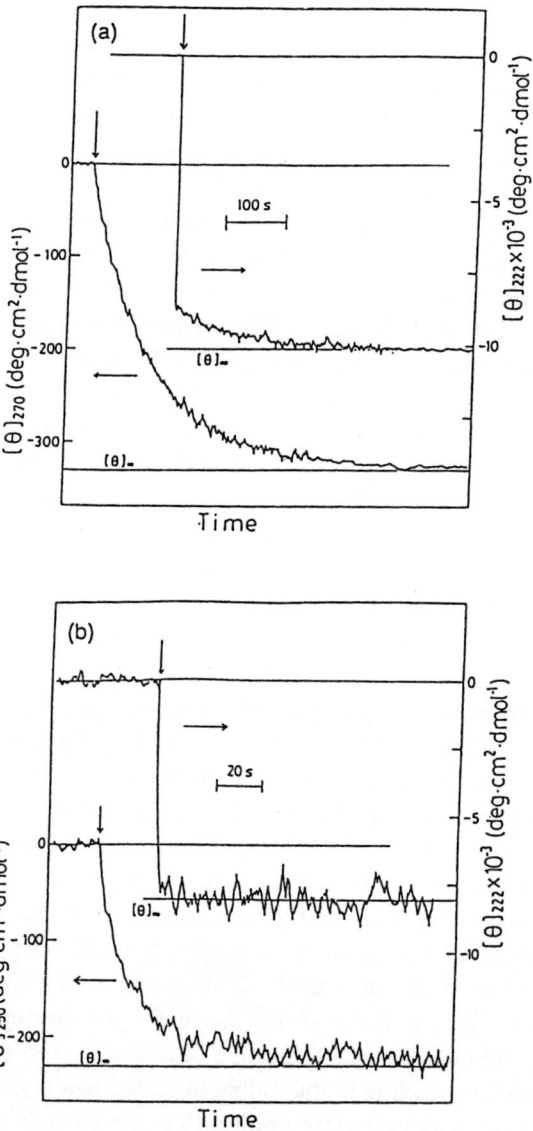

Fig. 1.11.   Kinetic measurements of refolding by means of CD spectra at 4.5°C. The refolding was initiated by a concentration jump of GdnHCl from 6.0 to 0.3M. (a) Apo-α-lactalbumin. (b) Lysozyme. Vertical arrows indicate the zero time at which refolding was initiated (reproduced from Ref. 59 with permission).

packing the secondary structures. The folding mechanism will thus be clarified as seen later. Ptitsyn[52] proposed the folding pathway which can be described in Fig. 1.12, where the intermediate state is composed of S and M.

The molten globule states in equilibrium were also investigated for α-lactalbumin with (holo) or without (apo) Ca^{2+} and lysozyme by Kuwajima *et al.*[59,60,61] They defined an empirical quantity of unfolding by

$$f = \frac{[\theta]_N - [\theta]}{[\theta]_N - [\theta]_U}, \tag{1.21}$$

where $[\theta]_N$ and $[\theta]_U$ are, respectively, the ellipticities of the native and the unfolded states. $1 - f$ is the degree of the formation of the secondary structures or that of the tertiary structures, depending on the choice of either far-ultra violet (222 nm) or near ultra violet (270 nm). Their results are shown in Figs. 1.13 and 1.9 for α-lactalbumin (apo- and holo-) and lysozyme, respectively. Figure 1.13 for apo-α-lactalbumin demonstrates that when the

Unfolded → Secondary Structures Only → Molten Globule → Native

(U)   (S)   (M)   (N)

Fig. 1.12.   Folding pathway.

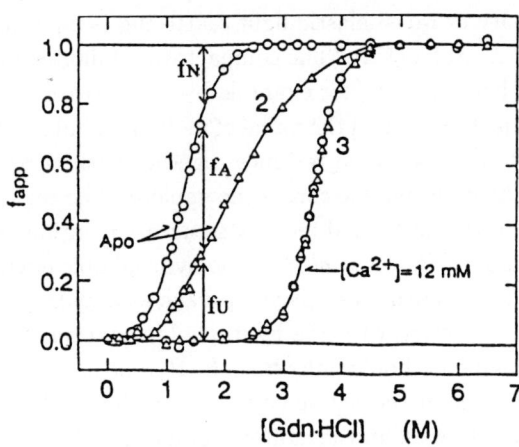

Fig. 1.13. f versus concentration of GdnHCl for α-lactalbumin. o at 270 nm and △ at 222 nm. Lines 1 and 2 refer to apo-α-lactalbumin and line 3 to holoprotein (reproduced from Ref. 59 with permission).

denaturant is diluted in the fully unfolded state, the secondary structure is formed gradually while the tertiary structures remain almost unfolded up to a certain concentration. On further decrease of the denaturant both the tertiary and the secondary structures are simultaneously folded. The intermediate structure thus observed is shown to be the same with the A state found at low pH first by Kronman *et al.*,[62] and then by Kuwajima *et al.*[59,63] These experiments supply an evidence of the presence of the intermediate states in equilibrium with the result to yield the three-state theory. In the Ca^{2+} binding holo-protein (Fig. 1.13) and lysozyme which cannot bind Ca^{2+} (Fig. 1.9), however, no intermediate structure is observed, and foldings (unfoldings as well) of the tertiary and the secondary structures proceed in a way as described by the two-state theory. It is interesting that α-lactalbumin and lysozyme are homologous proteins, and yet the the former has a stable molten globule in equilibrium state in apo-state, and not in holo-state, while the latter does not have stable molten globule. Further lysozyme also exhibits the kinetic molten globule state, as a first process of folding expected theoretically (see the case of chymotrypsin inhibitor 2 to be discussed later in this section).

The behaviors of the side chains can also be studied by NMR experiments. Ohgushi and Wada[64] showed that in horse cytochrome c the H^1-NMR spectrum of the aromatic region at the intermediate state has the features characteristic of the random coil state, but nevertheless the volume of the protein measured by intrinsic viscosity or quasi-elastic light scattering is only about 10% larger than that of the native structure (The compactness of molten globule state was also verified by other methods, for example by synchrotron small-angle X-ray scattering[65]). Thus, they coined the name of molten globule state. They meant by the term "molten" that some residues together with their side chains are mobile and thus the hydrophobic core is incomplete. The region of the molten hydrophobic core in the perturbed tertiary structure mentioned above is called the hydrophobic box by Baum *et al.*[66] or the hydrophobic environment by Mitaku *et al.*[67] The latter authors employed pyrene as a hydrophobic fluorescent probe to see the interaction with the hydrophobic environment in the molten globule state of bovine carbonic anhydrase B. Pyrene cannot bring about any anomalous fluorescence in the native as well as in the random coil state. The pyrene fluorescence observed in the molten globule state is attributed to the interaction with the molten hydrophobic core which enables pyrene to enter in it.

In the above description of the denaturation-renatureation phenomena, mention is often made of the two-state and three-state models. These states are also observed in the statistical mechanical simulation which will be discussed in Sec. 1.3.6. In the two-state theory no stable structure is observed other than U and N, and in the three-state theory another stable intermediate M can be observed. Further at the intermediate denaturing region two states or three states of the structures can be observed simultaneously and the thermodynamical relations such as Eq. (1.18) between the two states hold. The identity between the molten globule states found kinetically at the burst phase and those observed in equilibrium experiments were verified by Sugai, Kuwajima and their collaborators[57,69,70] for α-lactalbumin, lysozyme, and equine β-lactoglobulin. In some cases, for example in cytochrome c, the equilibrium molten globule states correspond not to the burst phase intermediates but to late folding intermediates.[71,72]

We now have a picture of the denaturing state and the molten globule state from the above observations. The molten globule at the equilibrium state has partially denatured secondary and tertiary structures in the sense that the formation and the packing of the secondary structures are incomplete (see Sec. 1.3.6). The kinetic molten globule obtained by denaturant-concentration jumps from random coil, i.e., completely unfolded state, has almost the same secondary structures with the native ones but partially denatured tertiary structure. The packing of the secondary structures in the molten globule is incomplete because of somewhat weakened and partially broken hydrophobic interactions. The remaining unbroken hydrophobic bindings are, however, still able to maintain a compact form similar to the native structure. This situation is appropriate to be called as molten hydrophobic core. The difference of the CD behaviors between apo-α-lactalbumin and lysozyme or Ca^{2+} containing holo-lactalbumin shown in Figs. 1.9 and 1.13 is due to the fact whether the conformation change is described by the two-state theory or not. The latter fact, in turn, comes from the stability of the intermediate structure, as mentioned already (see Sec. 1.3.7). An interesting example was given by Jackson and Fersht.[73] They showed that chymotrypsin inhibitor 2 exhibits the two-state transition in both equilibrium and kinetics, and no stable intermediate is observed. However, this does not invalidates the frame work represented by Fig. 1.12, because quick formation of secondary structures, whether they are stable or not, is the main event of protein folding, but further studies are required.

The molten globule state is usually transformed into the random coil state without significant change of heat content or specific heat as observed by

Ptitsyn.[52] Weak cooperativity in the unfolding of molten globules is also found in α-lactalbulin by Nozaka *et al.*[74] and Ikeguchi *et al.*[75,76] The transition between the random coil state and the molten globule is close to the secondary structure-coil transition, and not the first order transition, because the molten globule is mainly composed of rather short secondary structures, resulting in weak cooperativity, but the details will be different depending on the degree of the formation of the tertiary structure in the molten globule state. An example is shown in apo- and holo-α-lactalbulins,[77] where heat capaciy changes gradually with temperature in apo-α-lactalbulin, but it has an excess heat capacity in holo-α-lactalbulin. This is because the tertiary sructure is formed more in holo-α-lactalbulin than in apo-α-lactalbulin as verified by CD neasurements at 280 nm. Folding and unfolding, whether thermodynamical or kinetic, proceed by the formation or destruction of the secondary and tertiary structures, as found in the experiments shown in Figs. 1.9, 1.11 and 1.13.

1.3.5 *Graphical representations of protein structures*

Here a brief digression from the folding process will be made. In the studies of protein structures it is necessary to have some appropriate methods to visualize the structures. The conventional one is the two-dimensional projection of a three-dimensional structure. This gives us an intuitive picture of the structure, but it has the ambiguity due to the loss of the spatial relations. The latter

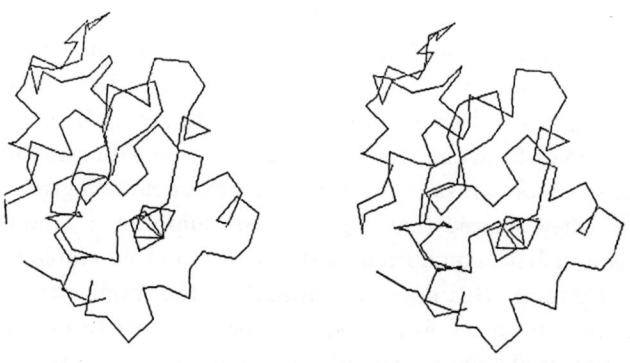

LYSOZYME(NATIVE)

Fig. 1.14. Stereopictures of hen egg-white lysozyme.

defect is remedied partially by the use of the stereopictures (Fig. 1.14), but still remains unsatisfactory in the quantitative aspects. Furthermore they depend on the direction of the projection plane.

A useful method is the distance map or the contact map.[13,78,79] An example is shown in Fig. 1.15 for hen egg-white lysozyme, where □ and ■ (blank and ×) indicate the pairs of residues whose α-carbons have the distance less (more) than 13 Å and ■ and × are the pairs of hydrophobic residues (in what follows

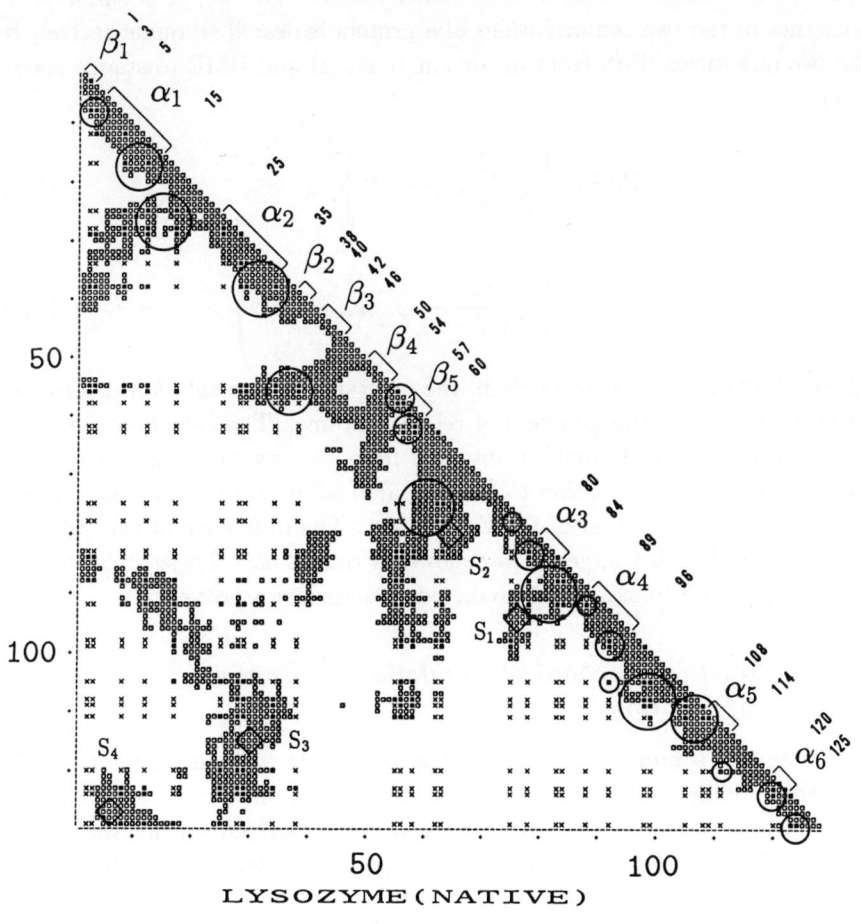

Fig. 1.15.   Distance map of lysozyme.

the same symbols will be used, if not stated otherwise). The disulfide bonds are
indicated by ◇. This distance map shows clearly the mutual spatial distances
of nearer residues and has an advantage that the figure is unique contrary to
the stereographic representation which is sight-direction dependent. It has,
however, a disadvantage that it loses the gross general view (see in the case
of myoglobin to be discussed later). The distance map when used for longer
space distances was shown to be useful for finding the modules in a protein by
M. Gō.[80] The distance map is usually drawn in a triangle, but when drawn in a
square, one sometimes calls it the distance matrix. The degree of similarity or
difference of the two conformations of a protein is described quantitatively by
the two quantities RMS (root mean square error) and DME (distance matrix
error):

$$\text{RMS} = \left\{ \frac{1}{N} \sum_{i}^{N} (r_i - r_i^c)^2 \right\}^{\frac{1}{2}}, \tag{1.22}$$

$$\text{DME} = \left\{ \frac{2}{N(N-1)} \sum_{i,j}^{N} (r_{ij} - r_{ij}^c)^2 \right\}^{\frac{1}{2}} \tag{1.23}$$

where r^c refers to the coordinate in the crystal or in the solution determined
by NMR and N is the numbers of relevant atoms. The calculation of RMS
requires that the conformation obtained from the crystal structure must be
placed near the conformation to be compared so as to minimize the value of
RMS, but the calculation of DME does not. The difference of the values of
RMS and DME is not large, as was shown by Sun *et al.*[51] These quantities are
usually employed to assess the predicted structure of a protein.

1.3.6   *Statistical mechanical simulation of protein conformation*

The folding of a protein takes place in a short time without misfoldings, and
thus the bindings between two residues existing in the actual structure can
be considered as never destroyed, once they are formed during the folding
process. An exception is the case of bovine pancreatic tripsin inhibitor (BPTI)
which will be discussed in Sec. 2.4. Now for a while without entering into
the problems of protein folding and the role of the hydrophobic interaction
mentioned already, we shall present an attempt to simulate the folding process

without misfolding. This approach was tried by Wako and Saitô.[81,82] in a one-dimensional lattice gas model. The details of their statistical mechanical calculations are not described here, but the main ideas and some of the results will be given here. As already discussed, the protein folding proceeds first by the birth of the embryos of the structure and then grow and coalesce into bigger structures. We describe this process by the island model, where an island is defined as the already constructed ordered structure either an α-helix, a β-structure, an antiparallel β-structure, or other intermediate structures. This model assumes that the folding process does not involve the formation of wrong structures, and therefore can be simulated by taking account of the interactions which exist in the tertiary structure of the native state. This idea was adopted by Ikegami[83] in his lattice theory of protein. Therefore this simulation by itself does not elucidate the folding mechanism, but will give some ideas on the protein folding.

In the statistical mechanical theory of the island model, the energy of a configuration of a protein is given by the following form

$$\varepsilon = \sum_{i=1}^{N} (P_i P_{i+1} \varepsilon_{i,i+1} + P_i P_{i+1} P_{i+2} \varepsilon_{i,i+2} + \cdots + P_i \cdots P_{i+n} \varepsilon_{i,i+n}). \quad (1.24)$$

In this formula N is the total number of residues of the protein, and n is the distance of the interaction which is taken 2 or 4 in case of β-strand or α-helix, but in proteins the possibility of $n = N$ must be considered. $\varepsilon_{j,j+k}$ stands for the interaction energy between the jth and the $(j + k)$th residues and is considered as a long-distance interaction for real proteins when k is large. $P_i = 1$ when the ith residue is in the native structure, and $P_i = 0$, otherwise. This formulation implies that the jth residue can interact with the $(j+k)$th residue only when the residues jth through the $(j+k)$th are all in the native structure. This is an important aspect of the island model. Besides this, the value of $\varepsilon_{j,j+k}$ is assumed negative if the three-dimensional distance between the jth and the $(j + k)$th residues in the native structure is small enough to allow interaction, and assumed zero, otherwise. The island model is shown to exhibit a first order phase transition in the limit of infinitely large values of N and n. Since for real proteins, however, N and n are finite, we always have a diffuse phase transition. The information of $\varepsilon_{i,k}$ can be obtained by the distance map, where we can put $\varepsilon_{i,k} = 0$ (the statistical weight $= 1$) if the (i, k) site is blank. Figures 1.16, 1.17 and 1.18 show the distance maps with cutoff distance of 7 Å of parvalbumin

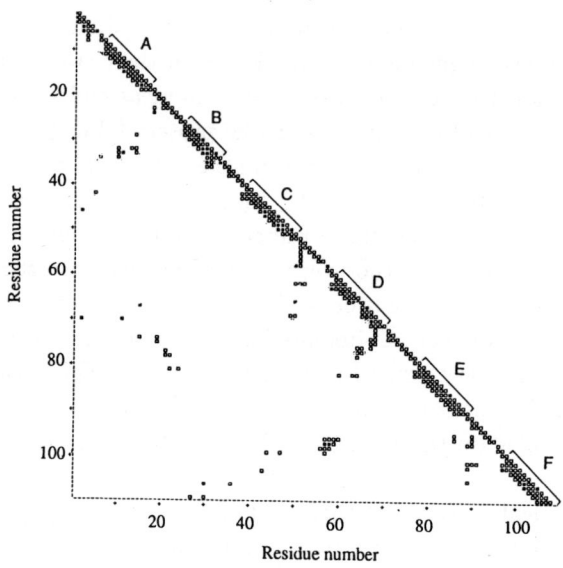

Fig. 1.16. Distance map of parvalbumin (reproduced from Ref. 82 with permission).

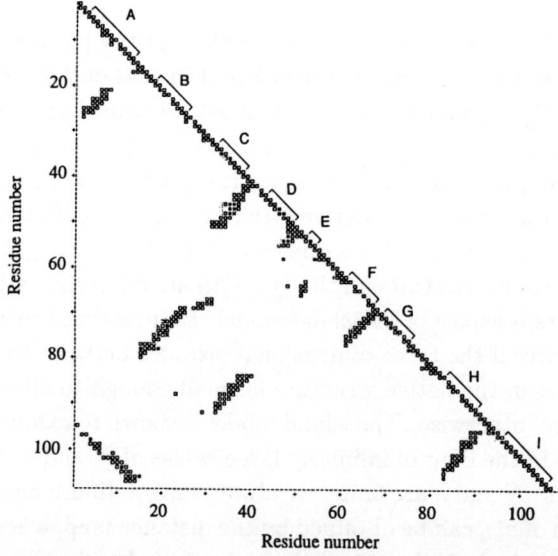

Fig. 1.17. Distance map of immunoglobulin (reproduced from Ref. 82 with permission).

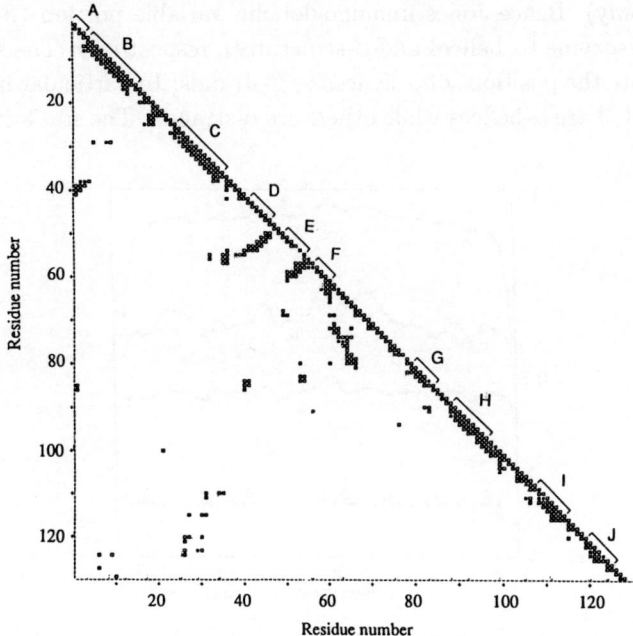

Fig. 1.18.   Distance map of lysozyme (reproduced from Ref. 82 with permission).

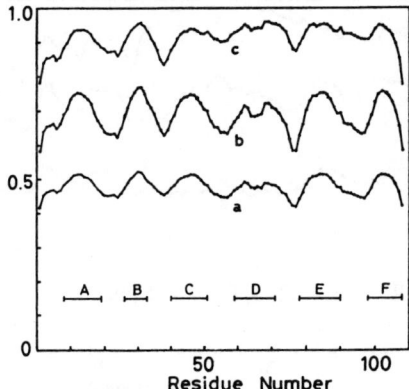

Fig. 1.19.   Degree of the formation of the structure of parvalbumin. (a) $z = 0.6$, (b) $z = 1.0$, (c) $z = 2.0$. Lines at the bottom indicate the location of the α-helices (reproduced from Ref. 82 with permission).

(α-helices only), Bence-Jones immunoglobulin variable portion (β-structures only) and lysozyme (α-helices and β-structures), respectively. The symbols A, B, ... indicate the positions of α-helices or β-strands. In particular in lysozyme B, C, G, H, I, J are α-helices while others are β-strands. The statistical weights

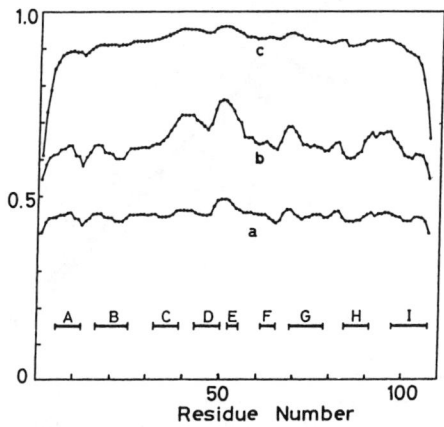

Fig. 1.20. Degree of the formation of the structure of immunoglobulin. (a) $z = 0.6$, (b) $z = 1.0$, (c) $z = 1.15$. Bold lines at the bottom indicate the location of the β-strands (reproduced from Ref. 82 with permission).

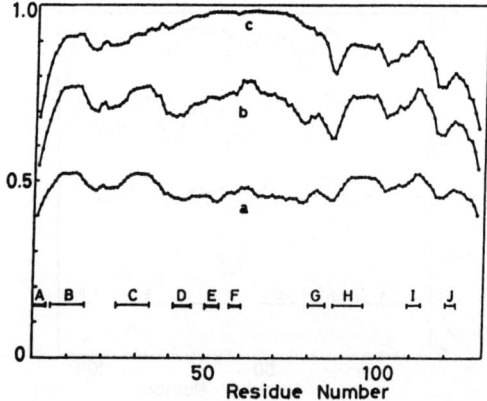

Fig. 1.21. Degree of the formation of the structure of lysozyme. (a) $z = 0.6$, (b) $z = 1.0$, (c) $z = 1.5$. A, D, E and F are β-strands and B, C, G, H, I and J are α-helices (reproduced from Ref. 82 with permission).

between two residues are put 1.2 or 1.0 if they are in contact or not, as can be determined by the distance map. Figures 1.19, 1.20 and 1.21 show the degrees of the formation of the structure of residues of parvalbumin, variable portion of Bence-Jones immunoglobulin and lysozyme, respectively. They also show that

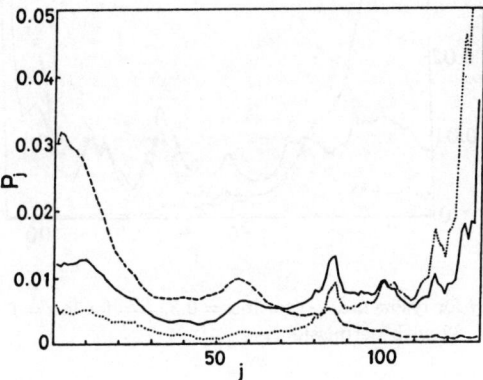

Fig. 1.22. p_j versus j for parvalbumin. $\cdots$ for $\ln z = 0.4$, —— for $\ln z = 0.9$, --- for $\ln z = 1$ (reproduced from Ref. 82 with permission).

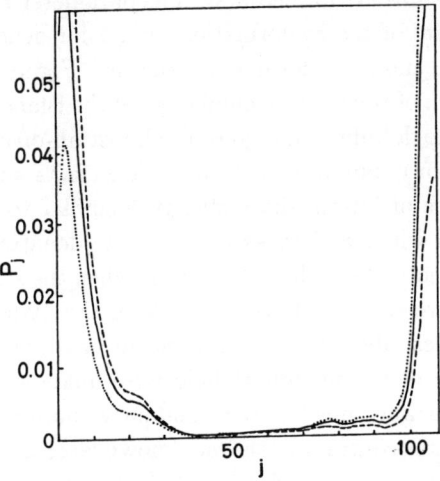

Fig. 1.23. p_j versus j for immunoglobulin. $\cdots$ for $\ln z = 0.05$, —— for $\ln z = 0.08$, --- for $\ln z = 0.11$ (reproduced from Ref. 82 with permission).

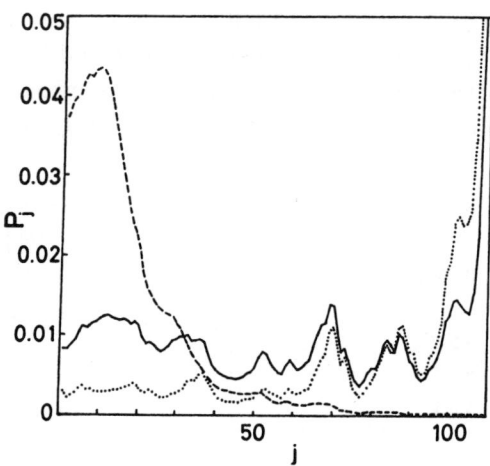

Fig. 1.24. p_j versus j for lysozyme. $\cdots$ for $\ln z = 0.3$, ——— for $\ln z = 0.6$, $---$ for $\ln z = 0.9$ (reproduced from Ref. 82 with permission).

the secondary structures are formed first. The parameter z is a quantity which can describe the state of the conformation: $z < 1$ for denatured states, $z \approx 1$ for transition region, and $z > 1$ for native structure. Figures 1.22, 1.23 and 1.24 show the distribution of the relative number p_j of the island of size j vs. j. One can see that immunoglobulin undergoes an almost all-or-none type transition (two-state description), but others do not. The facts suggest, interestingly enough, the presence of intermediate structure similar to the molten globule state in equilibrium when z changes from $z < 1$ (denatured state) to $z > 1$ (native state). It is noted here that at the time when the statistical mechanical theory presented here was developed the role of the hydrophobic interaction for protein folding described in Sec. 1.3.4 and to be discussed in Sec. 1.3.7, as well as the presence of the molten globule were unknown. In this sense the theory was too crude, and thus lysozyme and parvalbumin were shown to have intermediate structures contrary to the fact known later as absent (see Fig. 1.9). Probably the intermediate structures might disappear if proper accounts of hydrophobic interactions and chain entropy were taken into consideration (see the final part of the next section).

1.3.7 *Driving force for packing the secondary structures*

In the statistical mechanical theory of the island model developed in the above section, we have to know the tertiary structure. The tertiary structure itself should be predicted by the island model. To avoid the vicious circle, we consider among others the hydrophobic interaction which plays the essential role as discussed in Sec. 1.3.2. Many hydrophobic residues, however, are distributed along the protein chain and any pair of them has some possibility to

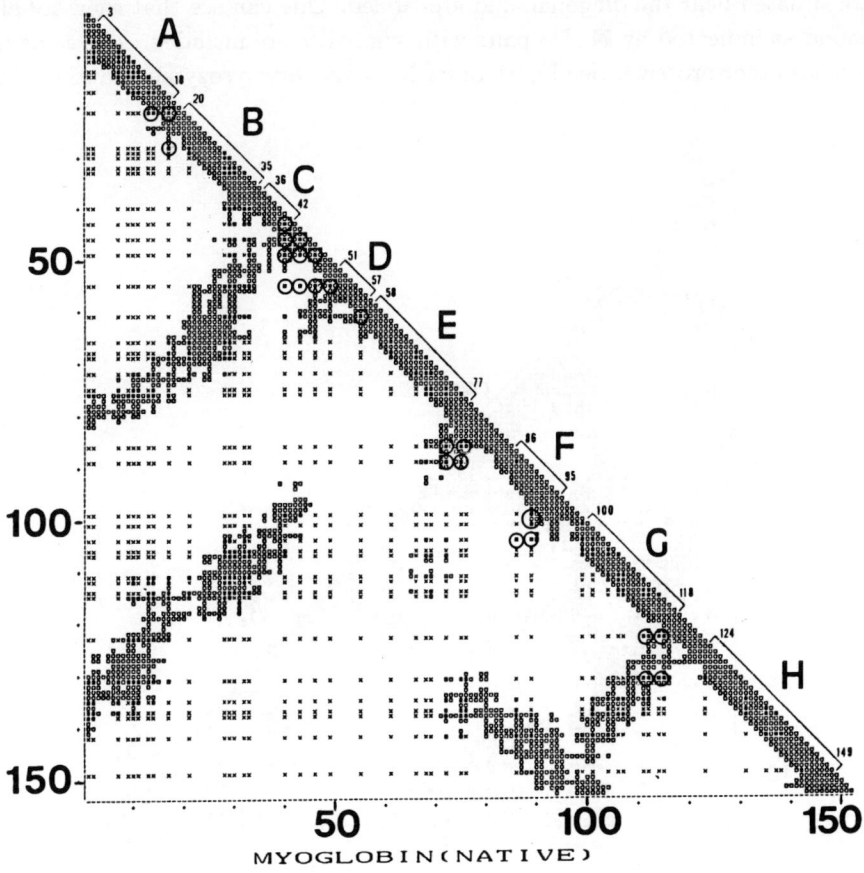

MYOGLOBIN(NATIVE)

Fig. 1.25. Distance map of sperm whale myoglobin (reproduced from Ref. 11 with permission).

make a hydrophobic interaction. But only specific pairs must be chosen to interact in order to make a definite structure. How this is done in real proteins can be guessed from the native tertiary structures which are supposed to be constructed without misfolding. The distance maps are most convenient for this purpose. The cutoff distance is chosen to be 13 Å, because the spatial distance between α-carbons of residues which are bound by hydrophobic interaction between their side chains is supposed less than this value. This value is found appropriate in many proteins. Figure 1.25 is the distance map of sperm whale myoglobin. The hydrophobic residue pairs at short distance are situated near the diagonal, and are circled. One can see that they are all bound as indicated by ■. No pairs with symbol × are included. The same is true for other proteins. See Fig. 1.15 for hen egg-white lysozyme and Fig. 1.26

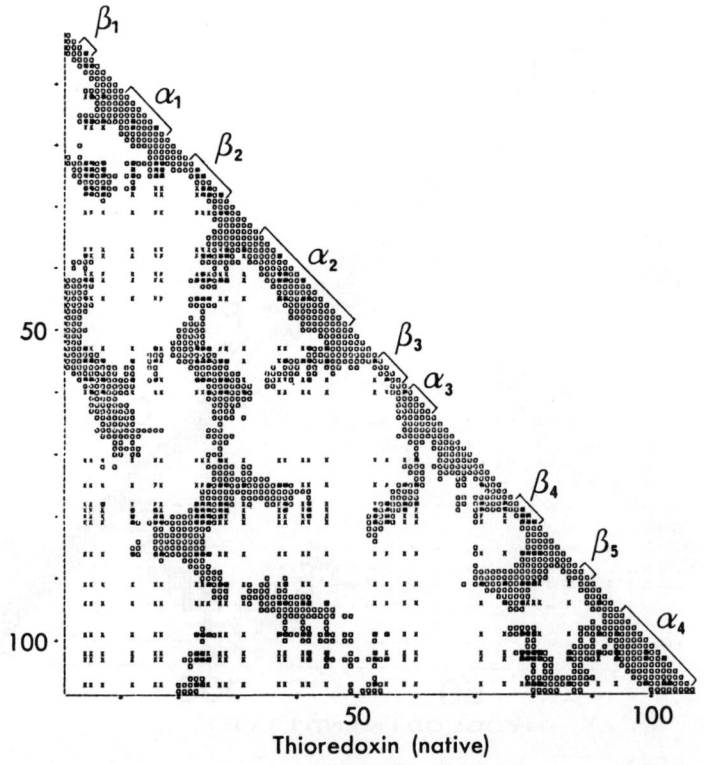

Fig. 1.26. Distance map of thioredoxin.

for flavodoxin. These observations imply that the circled hydrophobic pairs are essential for packing the secondary structures, and they can be specifically chosen from many possible pairs. In other words they are the pairs at shortest distance among them (as a matter of fact at medium distances on the chain, because usually hydrophobic residues are interspaced along the chain). To put it another way, they are made contact easily due to the long-range hydrophobic interactions (see Sec. 1.3.2) and quickly from kinetic point of view, with the least number of trials of changing the dihedral angles of the residues in between. This is what we have stated in Sec. 1.3.3.

We now reconsider this process from thermodynamics. The decrease of energy due to hydrophobic binding is much larger than the thermal energy of about 0.6 kcal/mol at ordinary temperature. In case when the distance of the two hydrophobic residues is short or medium, the decrease of the entropy of the chain between the bound residues is not large, and the free energy goes down. Thus the intermediate structure is stable, and the molten globule state is observed. Otherwise the intermediate structures are unstable, and one has two-state description.

To summarize, the hydrophobic interactions necessary for stabilizing the tertiary structures are assumed to be made between the residues at the shortest distance. In the case of BPTI, however, rearrangement of hydrophobic interactions takes place, contrary to the above consideration. But it is the only exception that has been confirmed up to now and will be discussed later (Sec. 2.4).

Chapter 2

Mechanism of Protein Folding

2.1 Island Model

In the above sections, we have elucidated several aspects of protein folding and proposed the island model for its mechanism. Up to here, one may consider that the island model looks similar to the conventional chain growth procedures,[84] but it is really quite different. This difference will be described below and the details of the model will be clarified by applying it to refolding of various proteins with known structures. The island model describes the protein folding process as follows. The initial random conformation of a polypeptide gradually changes when the conditions of environment such as temperature, concentration of the solvent components or pH is modified. The first event is the formation of the embryos of the ordered structure that are rather unstable, but when grown further they become stable to form α-helices or β-strands. Here is played the important role by the short-range and short-distance interactions such as van der Waals interaction and/or hydrogen bond. Then these secondary structures incorporate nearer residues or islands to become larger islands and finally reach the completely ordered structure, as shown in Fig. 2.1. In these processes, hydrophobic interactions between the hydrophobic residues at close distances are crucial for folding without trial and error. At the final stage the remote residues happen to make contact, which may slightly deform the already constructed structures.

In the following sections, we exclusively consider the problem of packing the known secondary structures that are given from the native structure or else, ignoring the problem of their formation. To implement the island model idea,

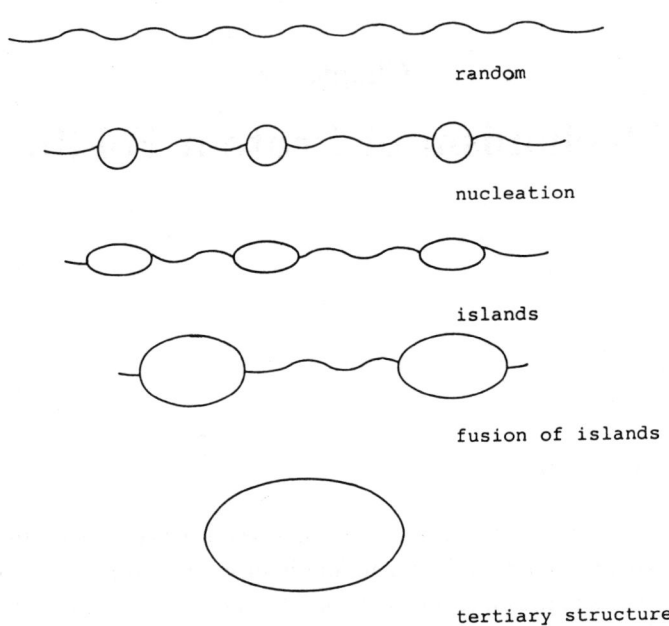

Fig. 2.1.   Schematic description of the island model.

we have to consider many aspects of protein structures, such as side chains and their flexibility, disulfide bonding, prosthetic groups, if any, and so on. Thus, we begin with a simple model proteins with only α-helices.

2.2 α-Helical Proteins

2.2.1   *Sperm whale myoglobin*

Sperm whale myoglobin[11] is composed of 153 amino acid residues and has eight α-helices (A, B, C,..., H) in the native structure (Fig. 1.25). Thus, the initial conformation is assumed to have eight α-helices of the native structure (strictly speaking, they must be of standard form), extended conformation in other regions and all the side chains of amino acid residues that are cut off beyond the β-carbons (except Gly). Figure 2.2 is the distance map with cutoff distances 13 Å of the initial conformation where the symbols are the same as in Figs. 1.15 and 1.25. We introduce the Lennard-Jones potential of

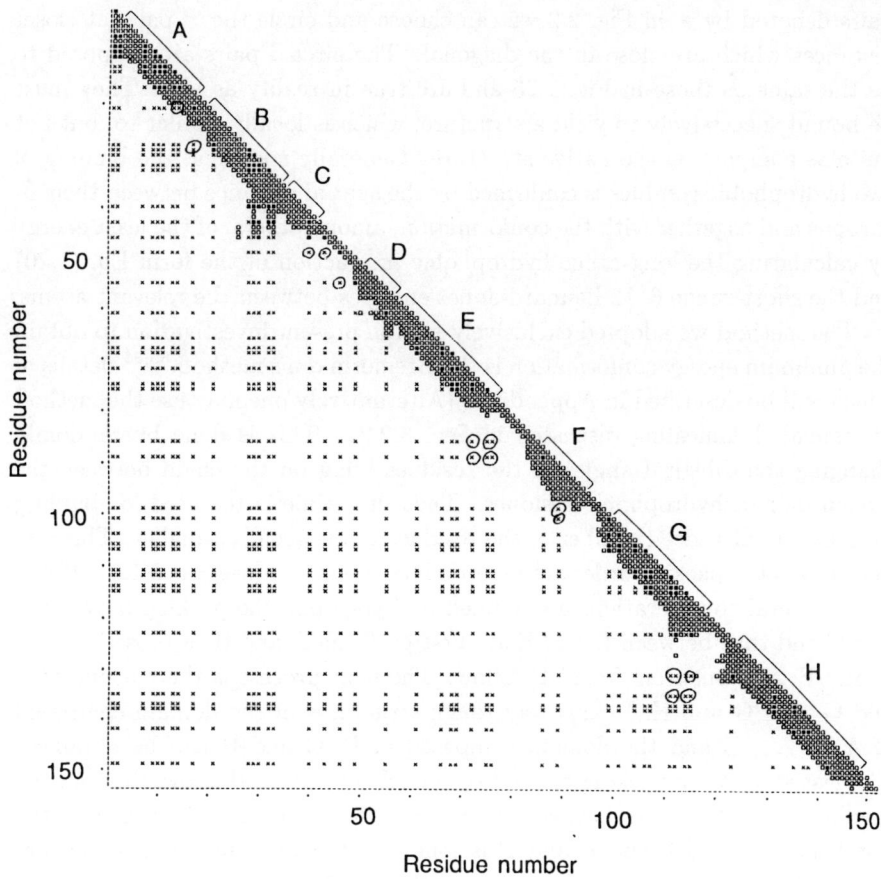

Fig. 2.2. Starting conformation for packing the eight helices.

6–12 type between nonbonded atoms (except H), and the hydrophobic interaction between the β-carbons of the hydrophobic residues. Other interactions
such as the hydrogen bonding in the regions outside the helices, electrostatic
interaction, and torsional energy of the dihedral angles are not taken into account. The parameters of the Lennard-Jones 6–12 potentials determined by
Momany *et al.*[84] are employed throughout our calculations. The hydrophobic interactions are assumed to be of the form (1.20) with $r_0 = 3$ Å for any
pair of hydrophobic residues. Looking at the distribution of the hydrophobic

pairs denoted by × in Fig. 2.2 we can choose and circle the × pairs at closer
distances which are close to the diagonal. The circled pairs are supposed to
be the same as those in Fig. 1.25 and are true in reality as well. They must
be bound successively to yield a structure, which is locally similar to, but not
quite as compact as the native structure. Generally speaking, the binding of
two hydrophobic residues is confirmed by the spatial distance between their β-
carbons and together with the conformation, among others, of the least energy
by calculating the long-range hydrophobic interaction of the form Eq. (1.20)
and the short-range 6–12 Lennard-Jones energies between the relevant atoms.

The method we adopted exclusively for the present investigation to obtain
the minimum energy conformation is the Bremermann's method,[85,86] details of
which will be described in Appendix C (Alternatively one may use the method
of simulated annealing discussed in Sec. 3.2.2). This is done by randomly
changing the dihedral angles of the residues lying on the chain between the
circled pair of hydrophobic residues. Thus, less time is required for binding
because the distance (number of the residues in between) is smaller. This fact
determines the packing order of the secondary structures (see Sec. 3.1.2). When
these general considerations are applied to myoglobin, the packing between B
and C and that between D and E are first performed, and then those between
A and B, and between C and D follow. The next process is the packing of F
and G, and G and H. In this way rough structures of the domain composed
of A, B,..., E and the domain composed of F, G and H can be obtained.
The last step is the packing of the two domains mentioned above through the
binding of circled pairs between E and F helices because the length of the
non-helical part between E and F is long but the hydrophobic residues are
scarce. The structure thus obtained is rather open and elongated. To reach a
more compact conformation we have to take into consideration of long-distance
hydrophobic pairs, marked × but not circled, successively from the nearer ones
to the remote ones. However, some of them cannot be bound because of the
steric hindrance arising from the structure already constructed, but others
will be bound successfully through the trials in reaching the minimum-energy
conformation by introducing interaction energies between relevant atoms.

The final conformation thus obtained through many processes of minimization is the refolded structure, that is presented in the distance map (Fig. 2.3).
One sees that Figs. 1.25 and 2.3 are quite similar in spite of the simple treatment, although the regions less than 13 Å are rather large compared with the
native structure. This is because the side chains are cut off at the β-carbons,

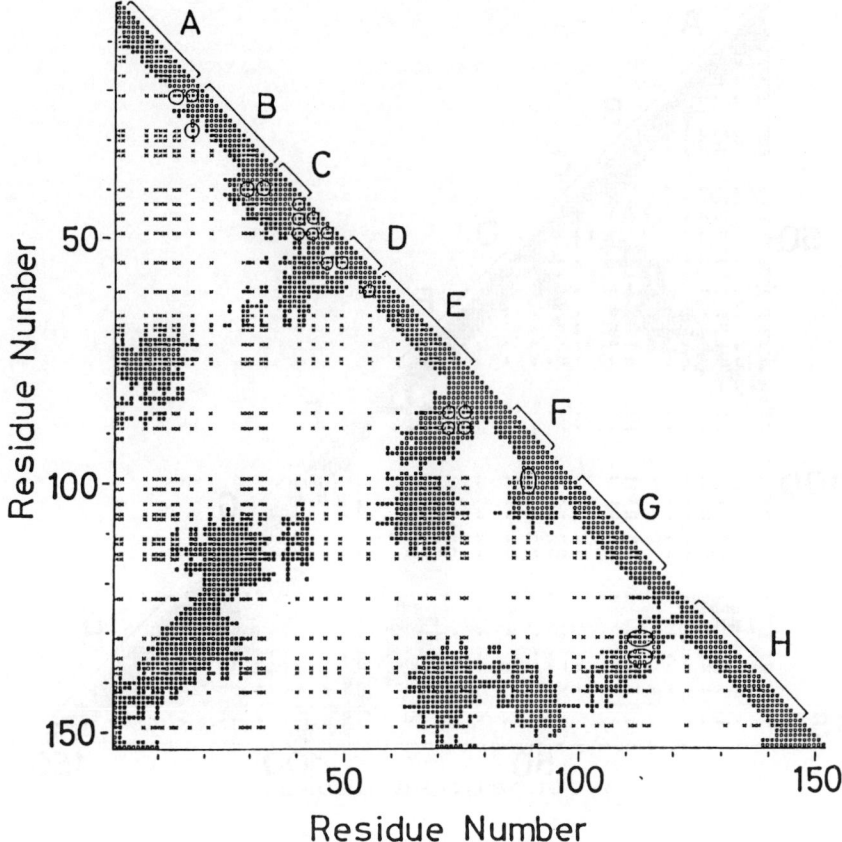

Fig. 2.3.   Refolded structure of myoglobin. Cutoff distance is 13 Å. All the side chains except Gly are assumed Ala type (reproduced from Ref. 11 with permission).

and the differences between the characteristics of side chains are not considered. Encouraged by this success despite the above mentioned crudeness of the specifications, we take into account the side chains, which are, however, assumed rigid spheres for simplicity. The geometrical parameters of the side chains are discussed in Ref. 11 and are summarized in the 3rd, 4th, and 5th columns of Table 1.6, where r_G is the distance from C^α to the center of gravity G of the side chain calculated from X-ray data and θ is the angle between the C^α–G and C^α–C^β directions. Their values in Table 1.6 are the averages

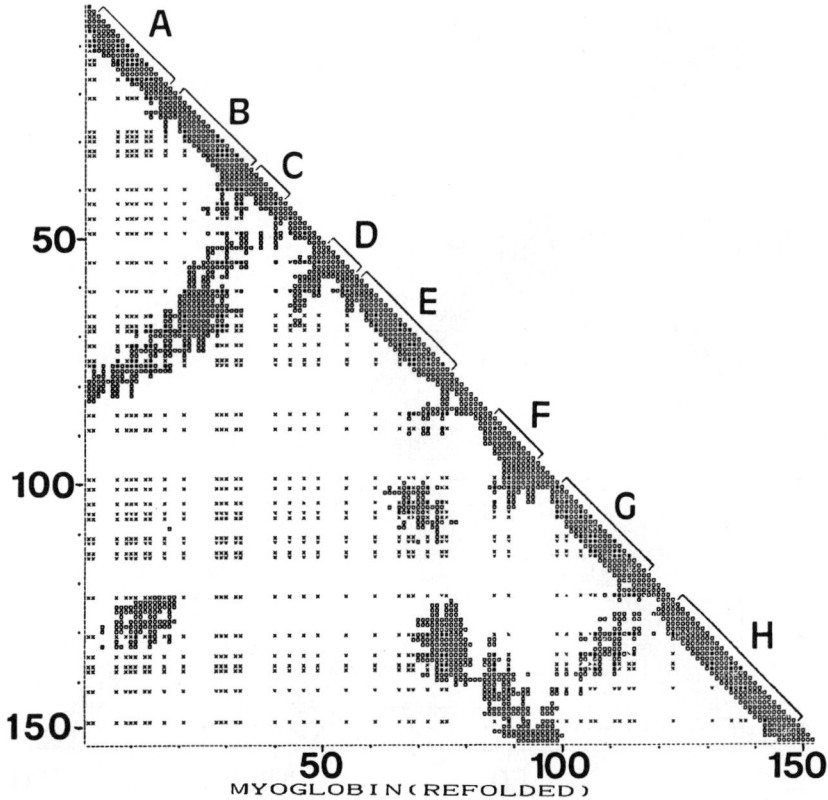

Fig. 2.4. Refolded structure of myoglobin without heme. Side chains are assumed as rigid
van der Waals spheres (reproduced from Ref. 11 with permission).

of 9 proteins.[a] Thus, in the new molecular model each amino acid residue is
replaced by a sphere of van der Waals radius located at the distance of r_G from
the α-carbon in the direction of β-carbon, ignoring the normally non zero value
of θ. We must use the value of the parameter r_0 in Eq. (1.20) which is equal
to the sum of the van der Waals radii of the relevant hydrophobic residues and

[a]These values of parameters are employed in the studies of various proteins, although they
are different from those of the real protein under consideration. Of course this procedure may
cause more or less a difference between the calculated and the native structures. However, at
the final stage of refinement the values in Table 1.6 will be changed to obtain more reasonable
structure (see Sec. 3.1.1).

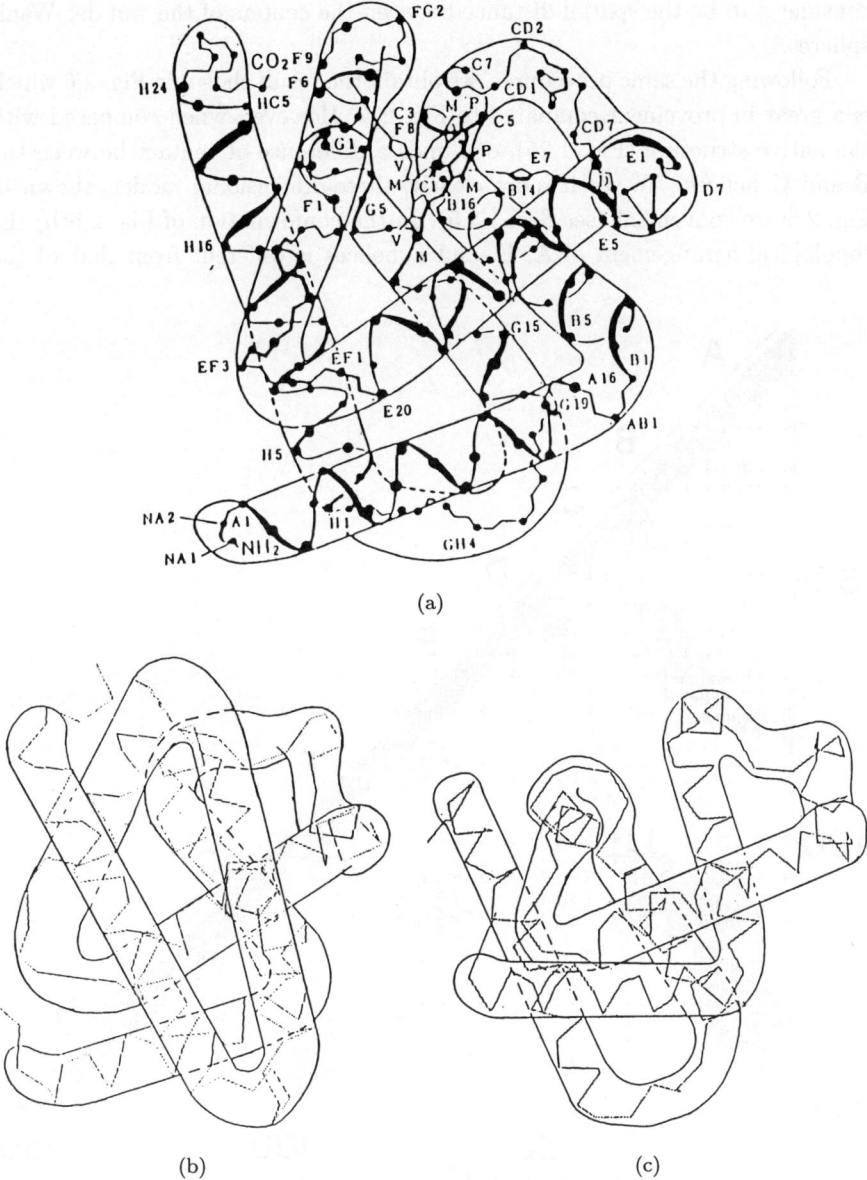

(a)

(b)   (c)

Fig. 2.5. Backbone structure of myoglobin (reproduced from Ref. 11 with permission).
(a) Native conformation (b) Refolded structure without heme (c) Refolded structure with
heme.

consider r to be the spatial distance between the centers of the van der Waals spheres.

Following the same procedures, we obtain the result shown in Fig. 2.6 which is a great improvement compared to Fig. 2.4. However, when compared with the native structure (Fig. 1.25), one finds a difference of contact between the B and G helices. To see it more clearly, three-dimensional models shown in Fig. 2.5 are convenient (see Sec. 1.3.5). In the conformation of Fig. 2.5(b) the topological arrangement of A, E, and F helices is different from that of the

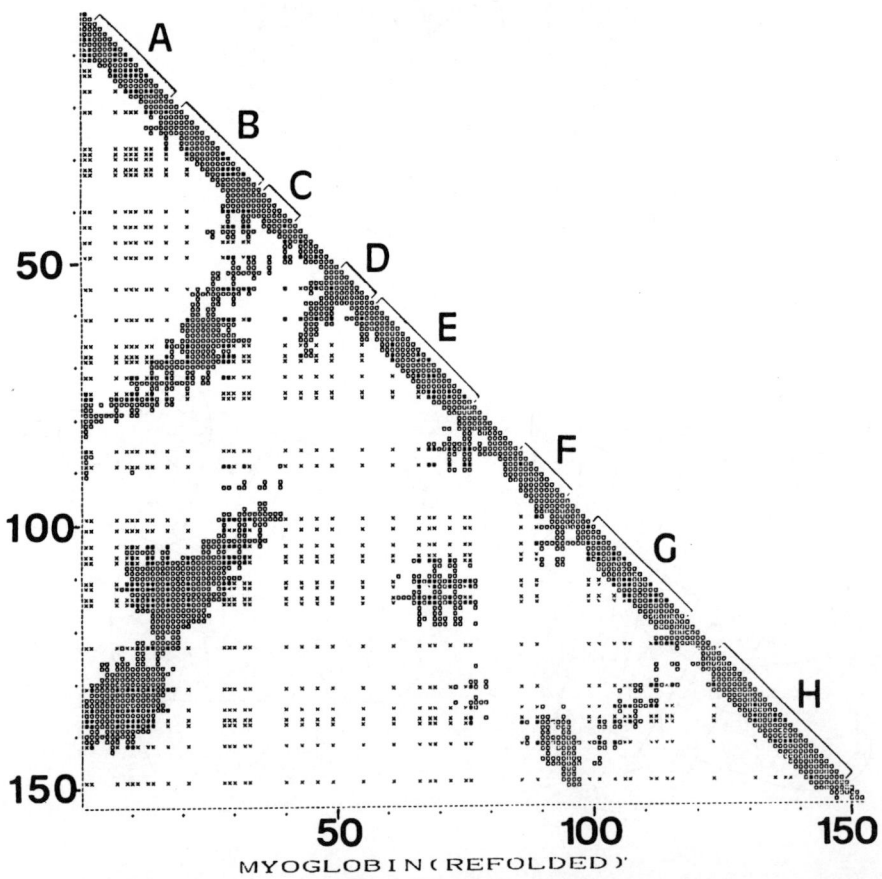

Fig. 2.6.   Refolded structure of myoglobin with rigid side chains and heme (reproduced from Ref. 11 with permission).

native one (Fig. 2.5(a)). In the native conformation the loop between F and G helices is bent in a complicated manner, and the two helices do not lie on the same plane (Fig. 2.5(a)). This is obviously due to the presence of the heme group. Thus, we introduced a heme group attached at histidine 93, represented by a sphere of radius 6 Å situated at a distance 6.4 Å from C^α in the direction of C^β. This is the distance between the C^α of His93 and the center of heme (Fe ion). Repeating the same procedures of packing described above, we could obtain the results presented in Figs. 2.5(c) and 2.6. Two domains consisting of the A, ..., E helices and the F, ..., H helices (with heme) were brought into

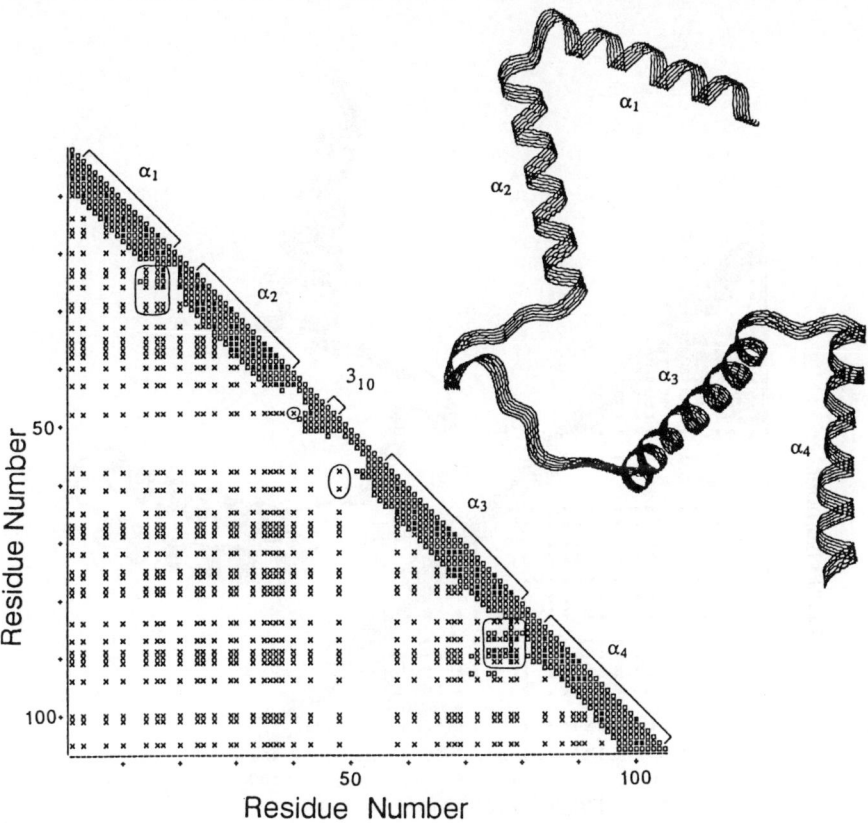

Fig. 2.7. Distance map and the ribbon structure of the initial extended conformation of cytochrome b_{562} (reproduced from Ref. 18 with permission).

contact by the hydrophobic interactions between the residues 70–75 and 85–90 by changing the dihedral angles of the residues in the loop region between the E and F helices at the last stage. The topologically correct arrangements of the helices were obtained and the distance map is quite similar to the native one.

2.2.2   *Cytochrome b_{562}*

Another simple model is cytochrome b_{562}[18] which is composed of four α-helices (3–19, 23–40, 56–80, 84–105), and a 3_{10}-helix (46–48) with a heme group attached to His102. We start with the initial conformation as shown in Fig. 2.7 where the four α-helices and a 3_{10}-helix are already formed and the same sym-

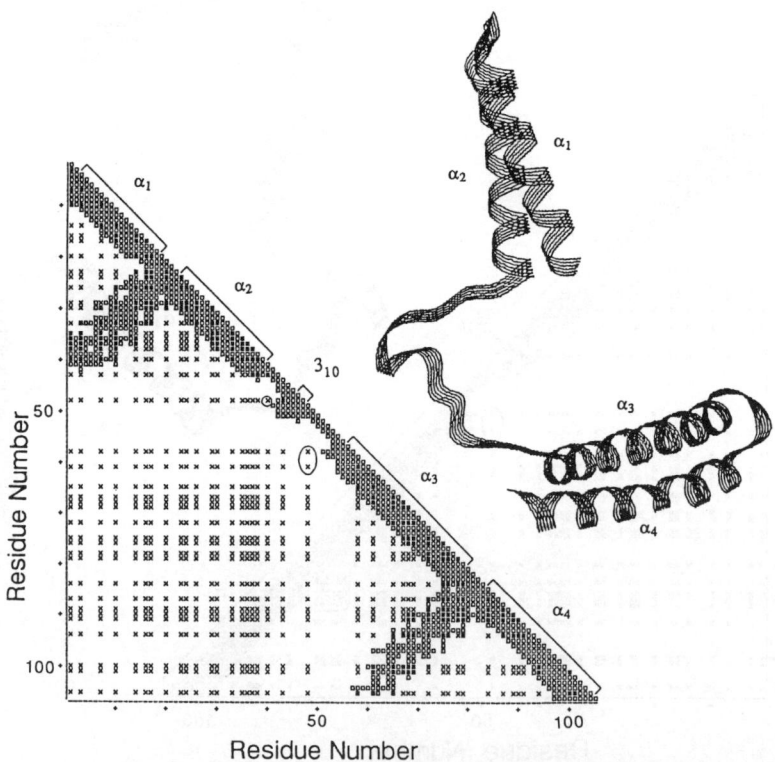

Fig. 2.8.   The distance map and the ribbon structure of cytochrome b_{562} at an intermediate structure (reproduced from Ref. 18 with permission).

bols are used as in previous figures. The important hydrophobic pairs are circled, but here Cys, Tyr, and Ala are included as weak hydrophobic residues, because, otherwise, hydrophobic residues are too scarce to pack the secondary structures. One can see a lot of hydrophobic pairs between α_1- and α_2-helices and between α_3- and α_4-helices. The helix pair of α_1 and α_2 and that of α_3 and α_4 are supposed to be formed first almost independently and simultaneously, yielding the conformation as shown in Fig. 2.8. These two helix pairs are then brought close to each other by changing the dihedral angles in the regions (41–45) and (49–55) to make the hydrophobic interaction between the circled pairs. In this process, four α-helices come close enough to make contact. Thus, the Lennard-Jones potentials and hydrophobic interactions are introduced between the residues of four α-helices which are mutually close in space. The effect of heme group has to be considered at this stage in the same manner as described in the case of sperm whale myoglobin. The refolded structure

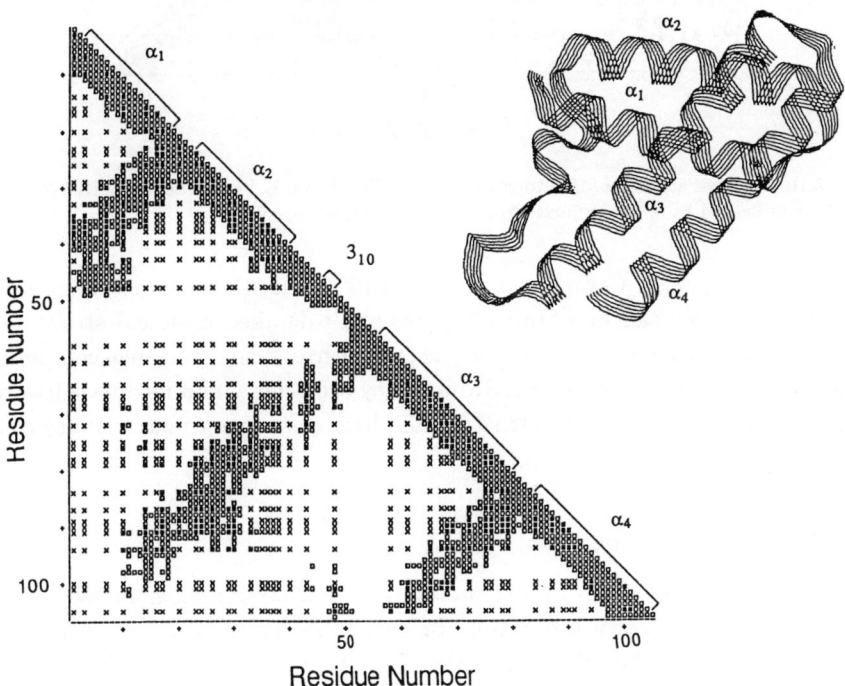

Fig. 2.9. Refolded structure of cytochrome b_{562} (reproduced from Ref. 18 with permission).

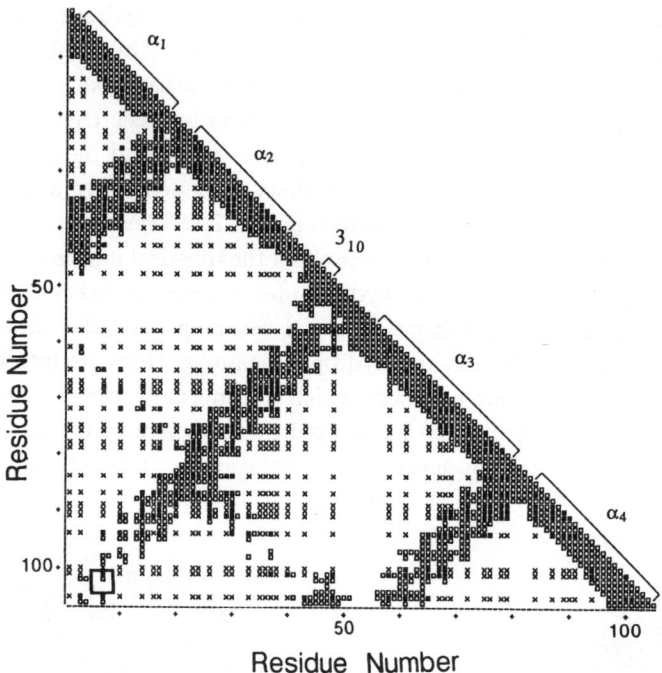

Fig. 2.10. Native structure of cytochrome b_{562}. The heme is bound between His102 and Met7 as indicated by 2 (reproduced from Ref. 18 with permission).

is given in Fig. 2.9. Compare it to Fig. 2.10 of the native structure. More refined structures can be obtained by restoring detailed chemical structures of amino acid residues. This can be accomplished easily by using computer programs in QUANTA and CHARMm. They are general and flexible software applications for modeling the structure and behavior of molecular systems (see Brooks *et al.*[87]).

2.2.3 *Sea hare myoglobin*

Our method is applicable to any protein of a low sequential homology with other proteins of known structure. In this subsection, we choose sea hare myoglobin[20] and explain the physicochemical origin of the structural similarity between sperm whale myoglobin mentioned already in Sec. 2.2.1 and sea hare myoglobin with a low sequential homology (28%).

```
  1'  VLSEGEWQLVLHVWAKVEADVAGHGQDILIRLFKSHPETLEKFDRFKHLKTEAEMKASED
      **..* .*. . ** * *. ...* *.*. **.. *.. . *. ** *. *..***
  1"  SLSAAEADLAGKSWAPVFANKNANGLDFLVALFEKFPDSANFFADFKG-KSVADIKASPK


 61'  LKKHGVTVLTALGAILKKKGHHEAELKPLAQSHATKHKIPIKYLEFISEAIIHVLHSRHP
      *.. . ..* *.... ... ......... . *..*
 60"  LRDVSSRIFTRLNEFV-NNAANAGKMSAMLSNFAKEHVGFGVGSAQFENVRSMFPGFVAS


121'  GDFGADAQGAMNKALELFRKDIAAKYKELGYQG


119"  VAAPPAGADAAWTKLFGLIIDALKAAGA
```

Fig. 2.11. Amino acid sequence comparison between sperm whale myoglobin and sea hare myoglobin by Genetyx-Mac. Identical residues between the two are indicated by (∗) and similar ones by (·).

Before entering into the refolding process, we compare the amino acid sequences and the secondary structures between these myoglobins. These sequences are tabulated in Fig. 2.11. Sea hare myoglobin is a 146-residue protein which consists of 7 α-helices A(4–19), B(21–35), C(51–55), D(59–76), E(81–97), F(102–118), G(126–143) and a 3_{10}-helix H(37–42) with a heme group attached to His95, while sperm whale myoglobin is a 153-residue protein which consists of 8 α-helices A(3–18), B(20–35), C(36–42), D(51–57), E(58–77), F(86–95), G(100–118) and H(124–149) with a heme group attached to His93. The positions of the secondary structures are similar to each other.

Next, we compare the distribution of hydrophobic residue pairs between these two myoglobins. The distance maps of the initial extended conformations are convenient for this comparison. Here, we consider only six strong hydrophobic residues. Figures 2.2 and 2.12 show a small similarity in the distribution of hydrophobic pairs between these myoglobins.

We simulated the refolding of sea hare myoglobin through the same process as sperm whale myoglobin, but we set the dihedral angles of secondary structures to the standard values ($\phi = -57°$, $\psi = -47°$) and at the final stage of simulation, we refined the conformation by generating side chain atoms with QUANTA and searching for the structure of minimum energy with CHARMm. In this way, we have reached a structure similar to sperm whale myoglobins as shown in Fig. 2.13 (compare it with Figs. 2.14 and 1.25). This indicates that independent of the sequence homology, secondary structures pack into

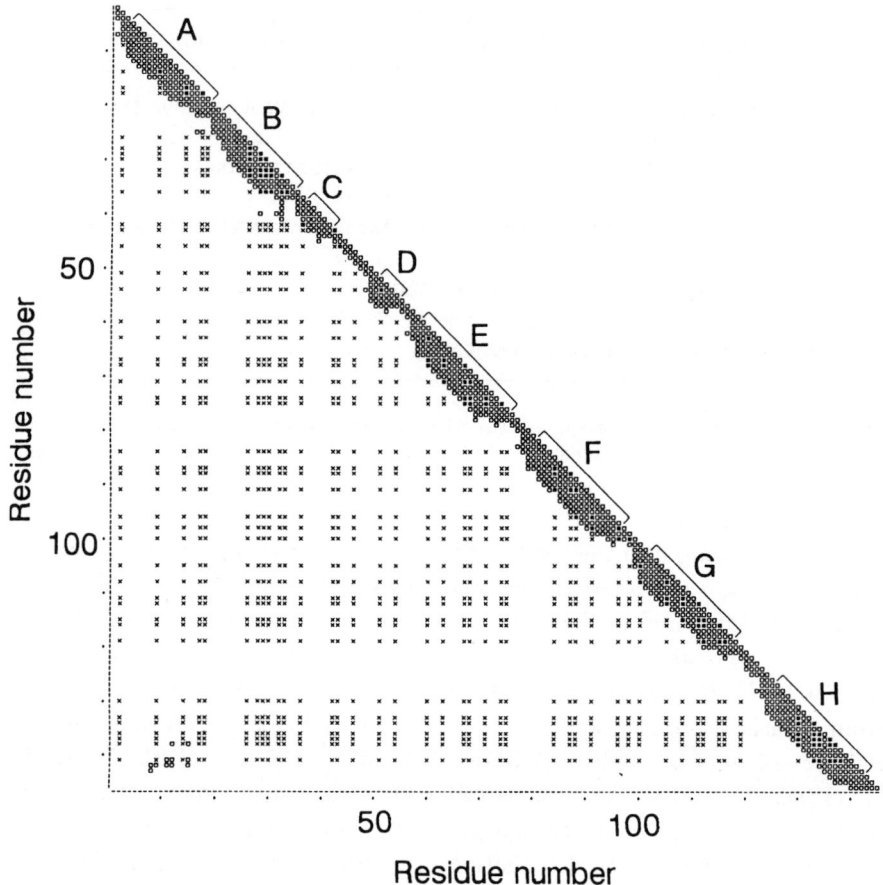

Fig. 2.12.   The initial extended conformation of sea hare myoglobin (reproduced from Ref. 20 with permission).

the tertiary structure by the hydrophobic interactions among amino acid pairs important to the local structure formation.

In the above treatment, we folded the initial conformation with α-helices of standard molecular parameters, in particular, with the dihedral angles of the values $\phi = -57°$ and $\psi = -47°$. Now, we tried to fold another initial conformation with dihedral angles of the native structure determined by the X-ray analysis. The procedures were, of course, the same but we reached a slightly

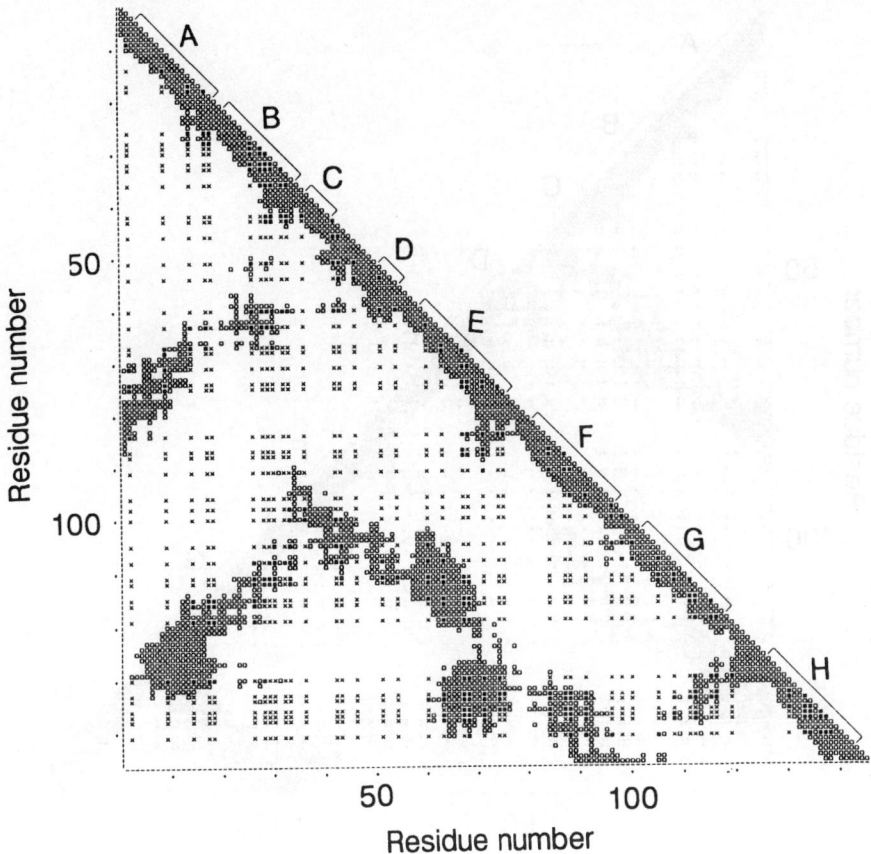

Fig. 2.13. The refolded structure of sea hare myoglobin. The distance matrix error (DME) over all C^α distances is 4.97 Å (reproduced from Ref. 20 with permission).

different conformation whose DME value was 10.2 Å, a value larger than the previous case shown in Fig. 2.13. This indicates that the initial standard helices are deformed during the folding process, but when the deformed helices are adopted at the initial conformation, they cannot fold properly (Sperm whale myoglobin was a lucky example). This implies that the folding order is essential to obtain the right conformation. Since the nascent polypeptide in biosynthesis is supposed to have the standard helices and fold into the native structure, it is reasonable to start with the initial conformation having the standard helices.

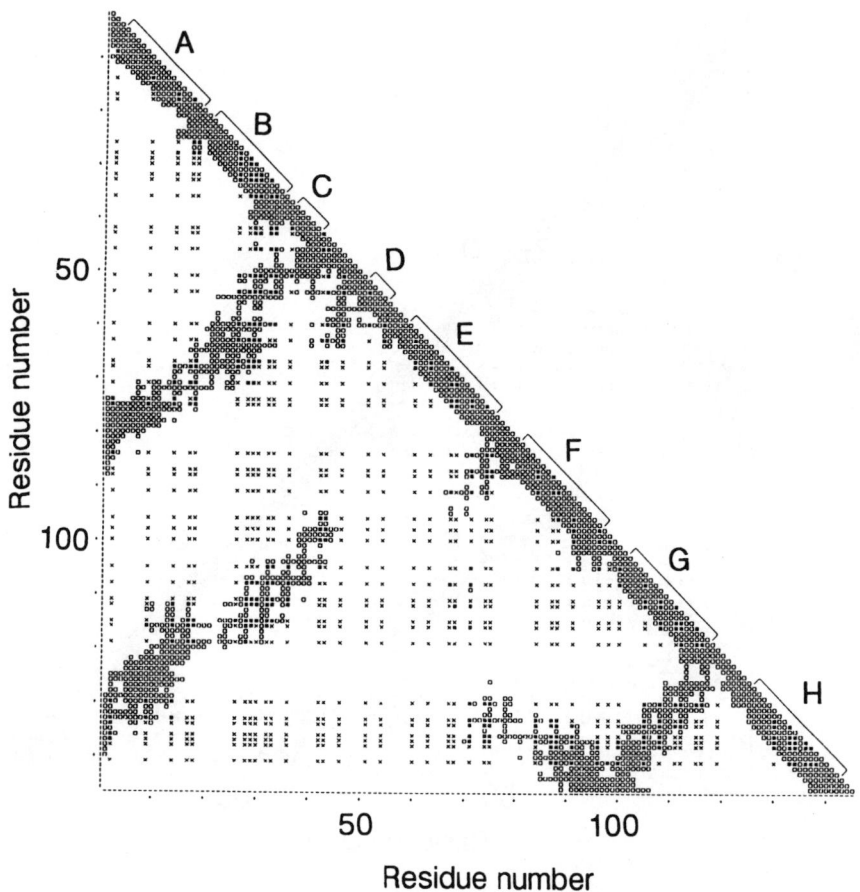

Fig. 2.14. The native structure of sea hare myoglobin (reproduced from Ref. 20 with permission).

2.3 Lysozyme and Phospholipase

Proteins are usually composed of α-helices, β-strands, and loops. These proteins are called $\alpha+\beta$ proteins. Also, many disulfide bonds fasten their chains to yield rigid structures. Lysozyme, phospholipase, and bovine pancreatic tripsin inhibitor (BPTI) are good examples. Before entering into this subject, a detour will be made on disulfide bonding.

2.3.1 *Disulfide bonding and lampshade*

The formation of a disulfide bond[15] is a process of oxidation. Two SH groups are required to come close to each other and some oxidants to come in between them. Thus, similarly to the case of hydrophobic interaction, the interaction between two S atoms is effectively long range. We may assume the energy of interaction to have the form:

$$E = \begin{cases} 20, & r < 3.2 \\ -10[1 - (r - 4.2)^2], & 3.2 \leq r < 5.0 \\ -3.6[1 - (r - 5.0)^2/25], & 5.0 \leq r < 10.0 \\ 0, & 10.0 \leq r \end{cases} \tag{2.1}$$

where energy E and distance r are given in units kcal/mol and Å, respectively. This expression is composed of two parts: a short-range chemical bond energy (S–S) and a long tail representing effective long-range interaction (Fig. 2.15). In the present calculation, the interaction energy (2.1) is introduced between the spheres representing the side chains of relevant cysteines. In order to make

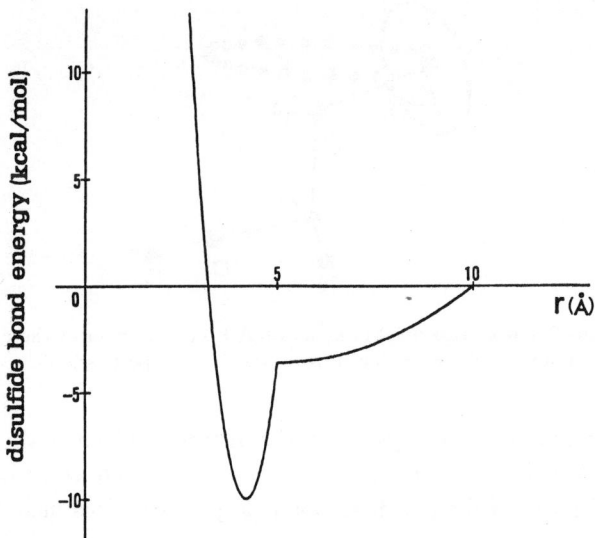

Fig. 2.15. Interaction energy in disulfide bonding (reproduced from Ref. 14 with permission).

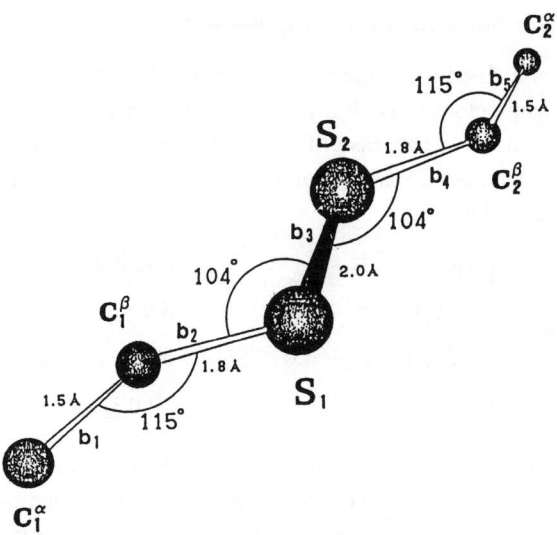

Fig. 2.16. Geometrical relation of a disulfide bond.

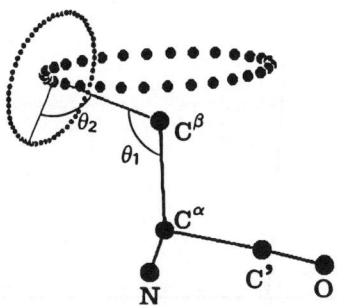

Fig. 2.17. Locus of H atom (assumed to lie at 1.0 Å from S atom or at the middle point of S–S bond) forms a lampshade (reproduced from Ref. 15 with permission).

a disulfide bond between two cysteines, the distance as well as the mutual orientation of the two cysteine residues must be appropriate so as to satisfy the directional property of the chemical bond (Fig. 2.16). This must be achieved by the folding mechanism described in the above sections, especially in Sec. 2.1 and is not by a random process. The geometrical relation required for bonding is easily visualized by the "lampshade" [15] shown in Fig. 2.17. Figures 2.18

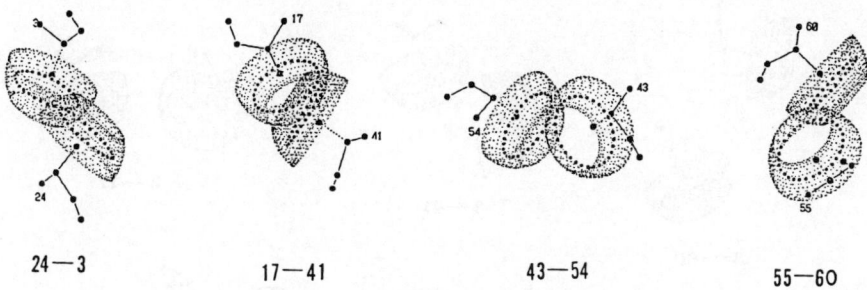

Fig. 2.18. Lampshades for bonded cysteines in erabutoxin (reproduced from Ref. 15 with permission).

and 2.19 show, respectively, the lampshades for bonded and nonbonded cysteine pairs in erabutoxin, which has 8 Cys's in 62 amino acid residues. The lampshades of bonded cysteines are face-to-face and crossed, while those of nonbonded pairs are not, though they are spatially proximate. Two exceptions 24–60 and 17–24 will be noted. They are unable to be bonded, because in the pair 24–60 Cys60 is already bonded with Cys55 at the earlier stages of folding and in the pair 17–24 Cys24 is bonded with Cys3. Thus, we have a criterion for disulfide bonding: simple translation due to changes of dihedral angles of rather remote residues can make two cysteine residues come close with their lampshades in the face-to-face relative position to make an intersection easily, without any steric hindrance of the main chain carrying the residues and others.

2.3.2 *Lysozyme*

The distance map of the native hen egg-white lysozyme is shown in Fig. 1.15. Three β-strands, β_3, β_4 and β_5 form a β-sheet (see Appendix D). The hydrophobic pairs located at the key positions for packing secondary structures are circled. Now we can start to refold the lysozyme[14] with the initial conformation having native α-helices and β-structures as shown in Fig. 2.20 where the loop parts are assumed to be extended β-like structures, and the side chains of amino acid residues are replaced by appropriate spheres of average radius situated at the average distances r_G from C^α in the direction of C^β as given in Table 1.6 and described in Sec. 2.2. In Fig. 2.20, hydrophobic pairs responsible for assembling the secondary structures are easily circled as indicated

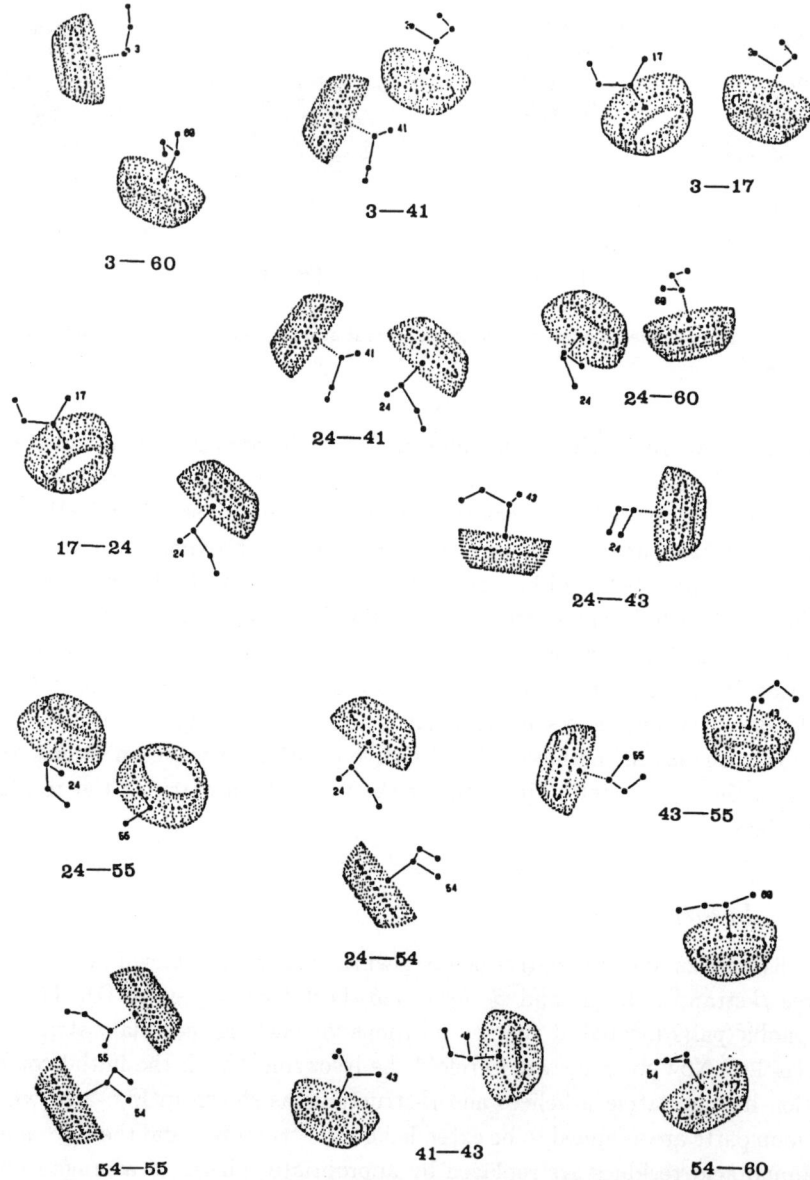

Fig. 2.19. Lampshades for nonbonded cysteine pairs in erabutoxin (reproduced from Ref. 15 with permission).

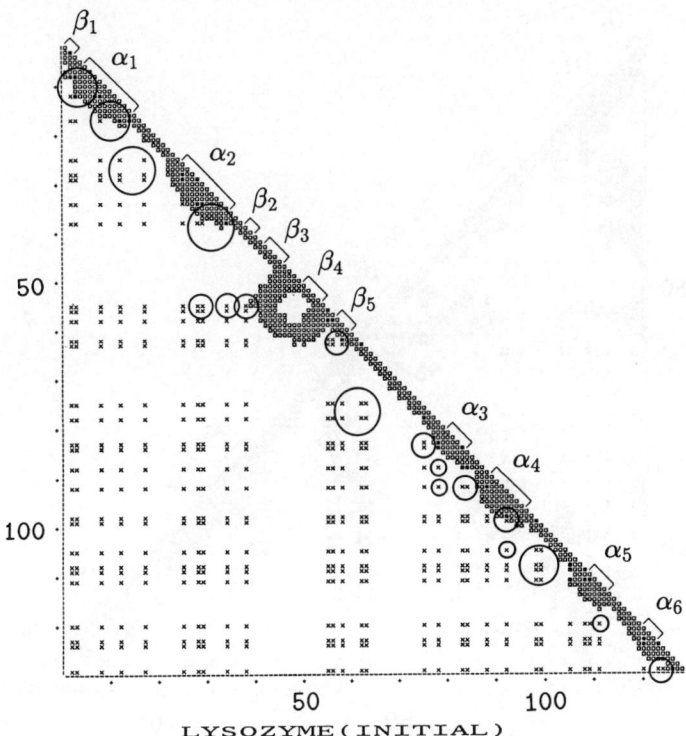

Fig. 2.20. Initial structure for refolding lysozyme (reproduced from Ref. 14 with permission).

there. They can be done without referring to the native structure and, more importantly, they are the same as in Fig. 1.15 as they should be.

By the same procedures employed in the case of myoglobin, we can refold the lysozyme.[14] A conformation obtained thus is shown in Fig. 2.21, where one sees that the short distance structures, or in other words local structures are almost the same with the native one, but long-distance structures are quite unsatisfactory. This is because the disulfide bonds are not introduced and thus the relevant cysteines are separated as can be seen in the Fig 2.21. The native structure has four disulfide bonds, S_1(Cys76–Cys94), S_2(Cys60–Cys64), S_3(Cys30–Cys115), and S_4(Cys5–Cys127). They are indicated by $\diamond$ in Figs. 1.15 and 2.21. Figure 2.21 shows that the spatial distances of C^α for the S_1 and S_2 pairs are both nearly 13 Å, but those for S_3 and S_4 pairs are

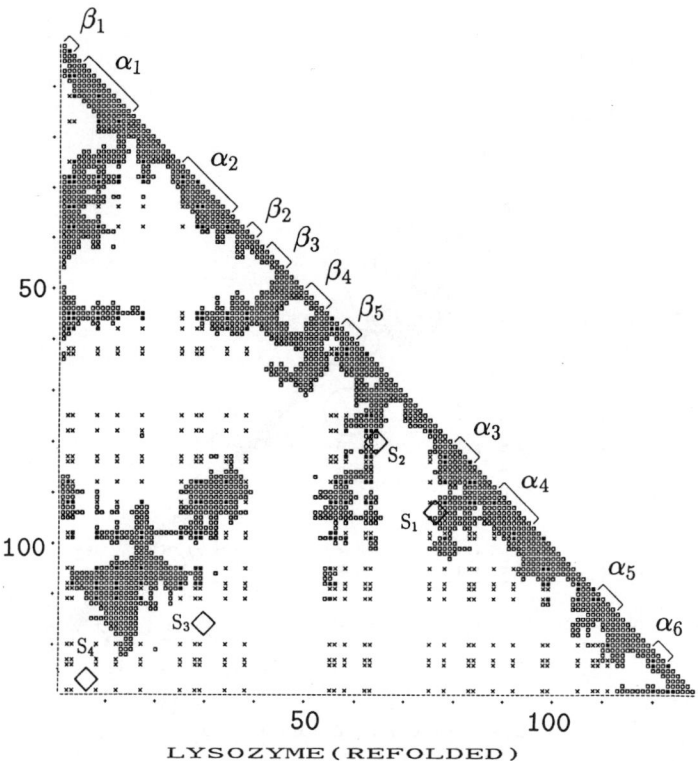

Fig. 2.21. Refolded structure of lysozyme without disulfide formation (reproduced from Ref. 14 with permission).

not. The disulfide bonds, S_1 and S_2 are supposed to be made easily. Among the possible pairs of cysteines, which are shown in Fig. 2.22 in the case of lysozyme, the appropriate pairs must be selected to yield the native structure. They are S_1 and S_2 in Fig. 2.22. This selection is carried out by means of "lampshades".[15] At the folding stage represented by Fig. 2.23 the mutual orientations of the lampshades of Cys76, 80, and 94 are shown in Fig. 2.24. It is easy to see that a disulfide bond is made between Cys76 and Cys94 (S_1) when the interaction energy (2.1) is introduced. The formation of S_2, which is approved by the lampshades of the pair, can be performed almost independently of the formation of S_1. When S_1 and S_2 are formed by introducing the interaction energy (2.1) at an early stage of folding, represented by Fig. 2.23, and

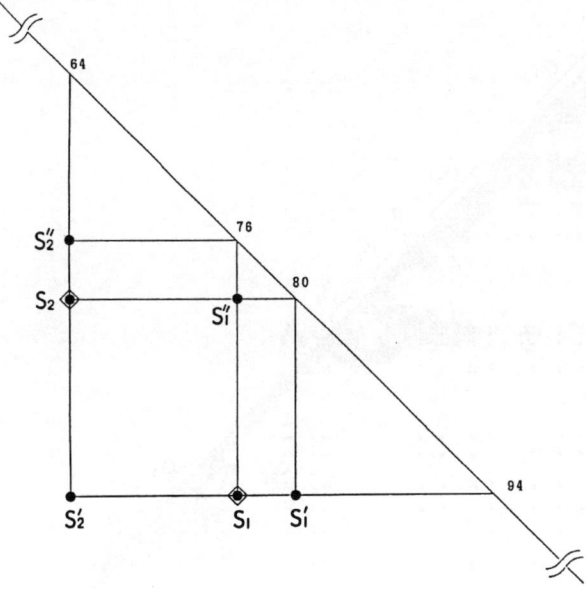

Fig. 2.22.   Possible Cys-Cys pairs in lysozyme (reproduced from Ref. 14 with permission).

further packing is performed, the structure becomes increasingly more similar to the native structure. However, disulfide bonds S_3 and S_4 cannot be formed still, as shown in Fig. 2.25. This is supposed to be due to the rigid side chains assumed in the present calculation and possibly due to the structure already determined in the final stage of folding, and will be remedied by using flexible side chains and by slightly deforming the final structure (see Sec. 3.1.1). This situation occurs in the case of phospholipase which is discussed in Sec. 2.3.3.

The sequence of the disulfide formation of S_1, S_2, S_3 and S_4 is in agreement with the experimental study on disulfide formation by Anderson and Wetlaufer.[88] Further computer simulations showed that unfolding of native lysozyme starts with the changes of the dihedral angles of the 100th $\sim$ 102nd residues. The simulation also show that the conformational fluctuations are large around the 20th, 100th, and 120th residues.[89,90] These facts are also consistent with the present conclusion. Acharya and Taniuchi,[91,92,93] however, showed through their extensive experimental studies that although the formation of disulfide bonds is not random, sometimes the pathway is flexible and an intermediate state with an open disulfide bond (S_3) between Cys30 and

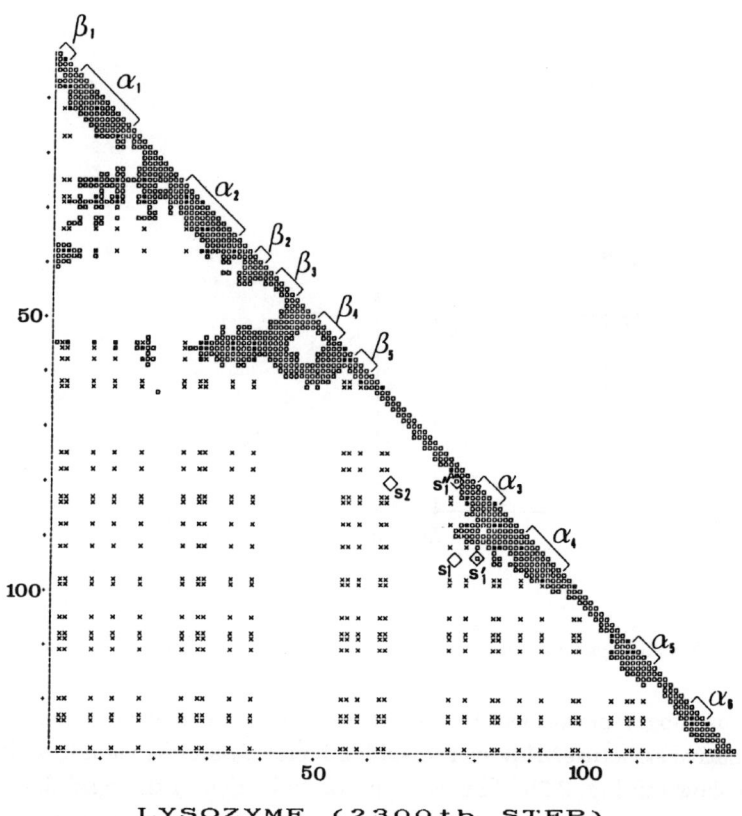

Fig. 2.23. An intermediate refolded structure of lysozyme (2300th step without S–S) (reproduced from Ref. 14 with permission).

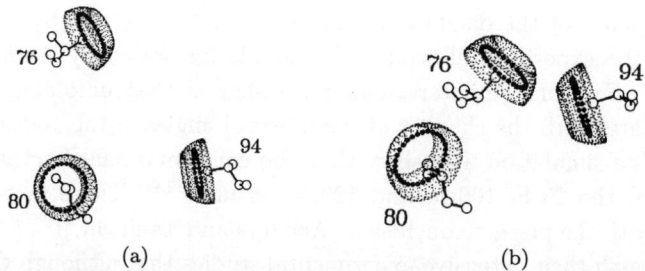

Fig. 2.24. Lampshades for Cys76, 80 and 94 viewed from different directions (reproduced from Ref. 14 with permission).

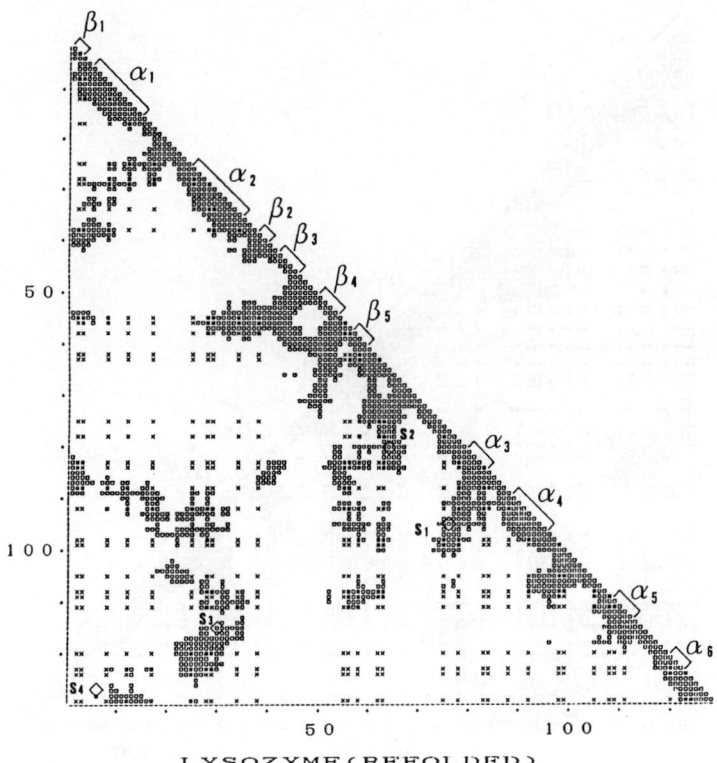

Fig. 2.25.   Refolded structure of lysozyme with disulfide bonds S_1 and S_2 (reproduced from Ref. 14 with permission).

Cys115 may or may not be observed. These can be interpreted as follows. The hydrophobic interactions bring the Cys of S_1 and those of S_2 close enough to be paired, but the formation of either S_1 or S_2 is sufficient to the formation of S_3 and S_4. The formation of S_3 is not necessary for the formation of S_4, because at this stage of folding the formation of S_4 is possible before the formation of S_3.

2.3.3   *Phospholipase*

Bovine pancreatic phospholipase[14] has 5 α-helices and 2 β-strands and 7 disulfide bonds. The native structure is illustrated in Fig. 2.26, where Trp, Ile, Leu,

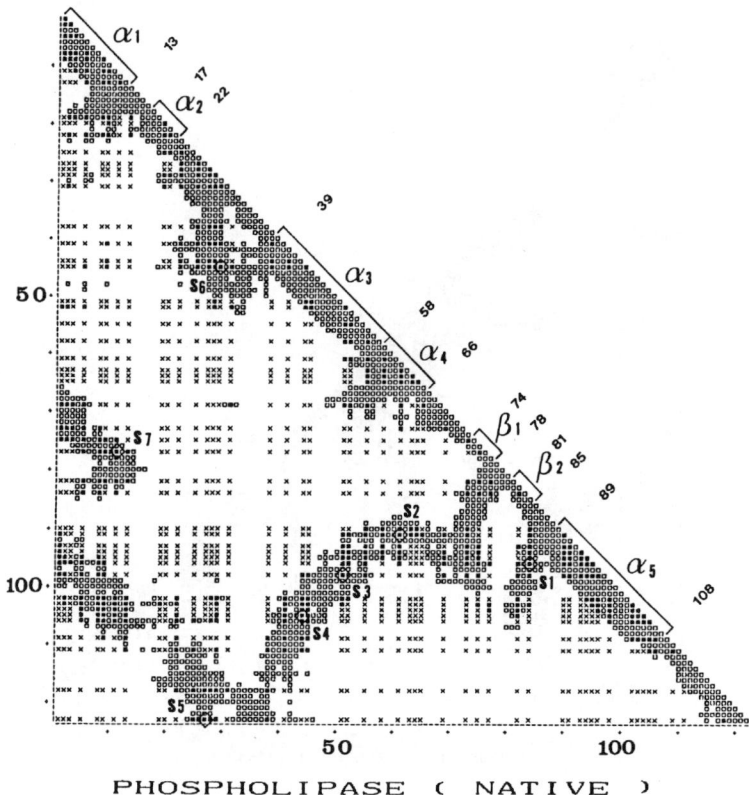

Fig. 2.26.   Distance map of the native structure of phospholipase (reproduced from Ref. 14 with permission).

Val, Phe, Met, and Cys are assigned as hydrophobic residues. Two β-strands form an antiparallel β-structure (see Appendix D). The seven disulfide bonds are called S_1(Cys84–Cys96), S_2(Cys61–Cys91), S_3(Cys51–Cys98), S_4(Cys44–Cys105), S_5(Cys29–Cys45), S_6(Cys27–Cys123), and S_7(Cys11–Cys77). The initial conformation we adopted is shown in Fig. 2.27, where the hydrophobic pairs responsible for packing can be circled without prior knowledge of the tertiary structure. We take other hydrophobic pairs into account when they come close together as folding proceeds. Some of them may not be bound due to steric effect. By looking at the distribution of pairs marked by circled $\times$ in Fig. 2.27, it is suggested that the parts $(\alpha_1 - \alpha_2)$, $(\alpha_3 - \alpha_4)$, and $(\beta_1 - \beta_2 - \alpha_5)$

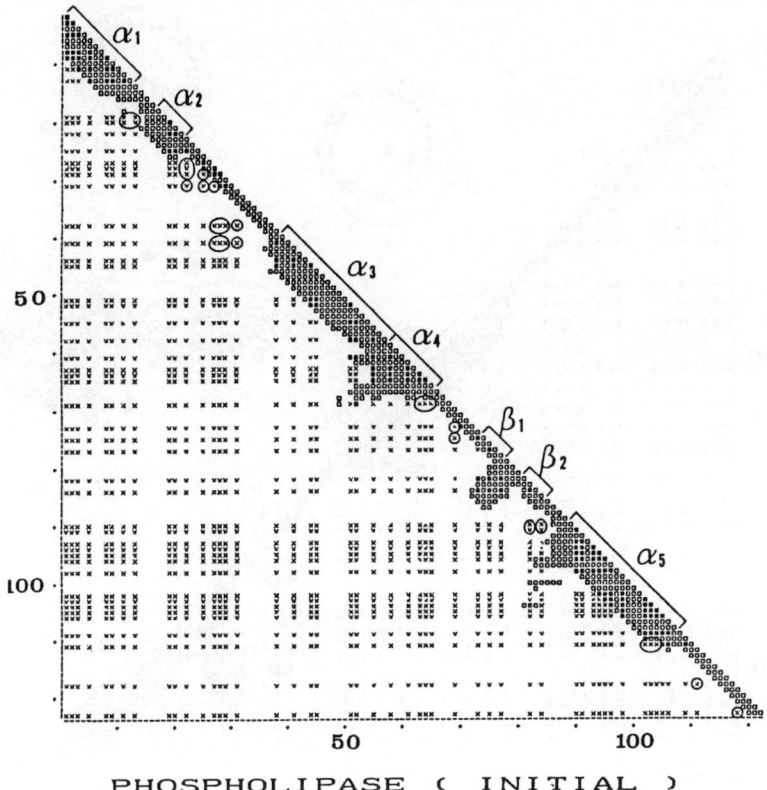

Fig. 2.27. Initial structure for folding phospholipase (reproduced from Ref. 14 with permission).

can rather quickly form three domains, because the parts between α_2 and α_3, and between α_4 and β_1 are long and yet lack enough hydrophobic pairs, with the result that the folding of these parts take a longer time. Consequently, we first make the bent structure of α_3 and α_4, which is obtained by changing the dihedral angles of the 58th residue, and then fold the parts $(\beta_1 - \beta_2)$ and α_5. The structure thus obtained is shown in Fig. 2.28, where Cys84 and Cys96 are sufficiently close to form the disulfide bond S_1, and the mutual relation among the lampshades of Cys84, 91, 96, and 98 is also shown. One can see that only the pair $S_1(84$–$96)$ is formed. After bonding the pair S_1, the lampshades of Cys61, 91, and 98 are examined to confirm the formation of S_2 (Fig. 2.29). In

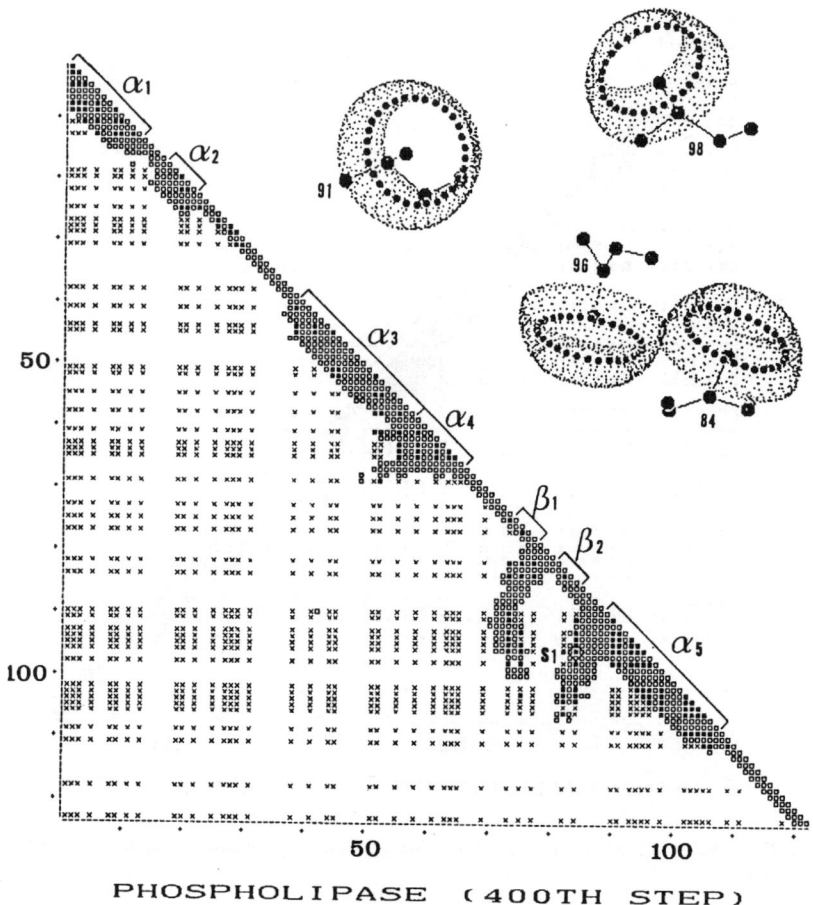

PHOSPHOLIPASE (400TH STEP)

Fig. 2.28. Partially refolded structure of phospholipase (reproduced from Ref. 14 with permission).

this way the pairs S_3 and S_4 can be formed successively. The last step is to search for the contact between the domain $\alpha_1 - \alpha_2$ and the domain $\alpha_3 - \alpha_5$. This is performed by changing the dihedral angles of the residues of the part between α_2 and α_3. The pairs S_6 and S_7 are bonded to yield the final result shown in Fig. 2.30 where, nevertheless, the pair S_5 is still not bonded. This is a similar situation as in lysozyme.

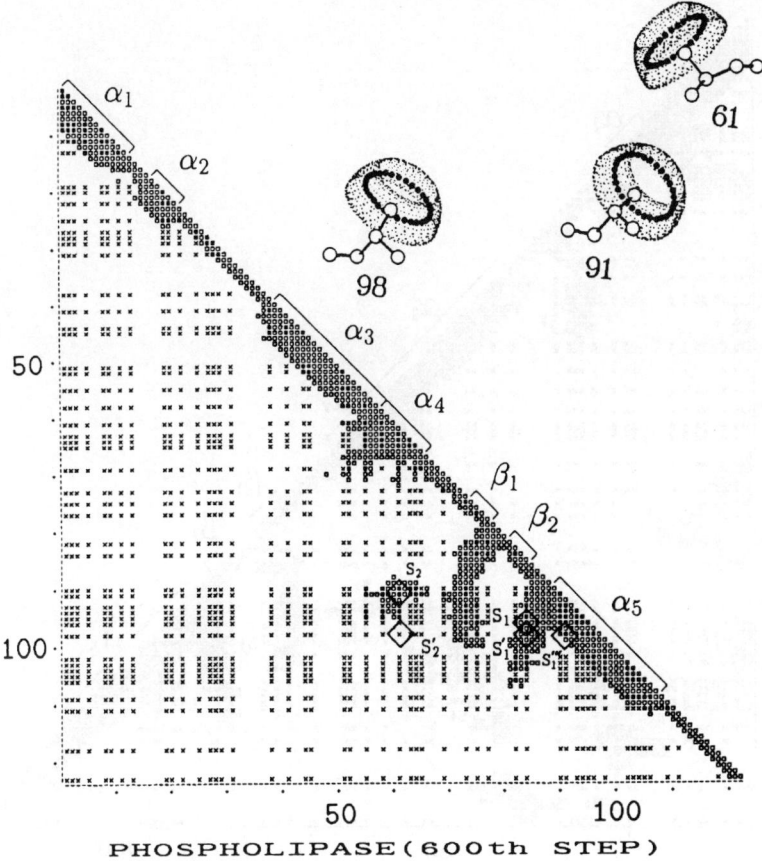

Fig. 2.29. Distance map and lampshades of Cys61, 91 and 98 at 600th step (reproduced from Ref. 14 with permission).

2.4   Bovine Pancreatic Trypsin Inhibitor

Bovine pancreatic trypsin inhibitor (BPTI)[16] is composed of 58 amino acid residues and has 3 disulfide bonds, an antiparallel β-structure, and a helix as shown in Fig. 2.31, where disulfide bonds are designated by $\diamond$. Important hydrophobic pairs close to the diagonal are regarded as the key residues that may connect the neighboring secondary structures. They are circled and marked as (a), (b), (c), (d), and (e). Inspecting them, one will find that the pairs in group

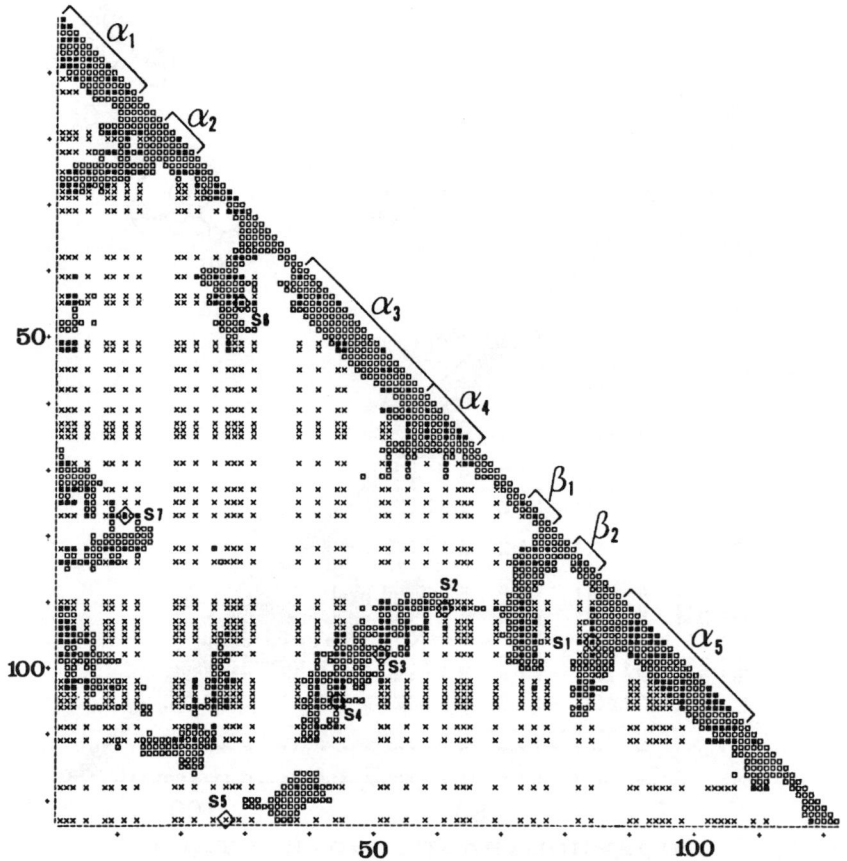

Fig. 2.30. Refolded structure of phospholipase at 2000th step (reproduced from Ref. 14 with permission).

(a) remain unbound in the native structure, contrary to the expectation of the island model. However, we are led to proceed to fold the initial conformation in accordance with the method of the island model.

Before discussing the results, it will be appropriate to describe the extensive experimental investigations of BPTI renaturation by Creighton and his collaborators started in 1975.[94-100] They showed that the reduced BPTI without disulfide bonds was refolded into the native structure via intermediate structures with nonnative disulfide bonds, as described in Fig. 2.32(a). The

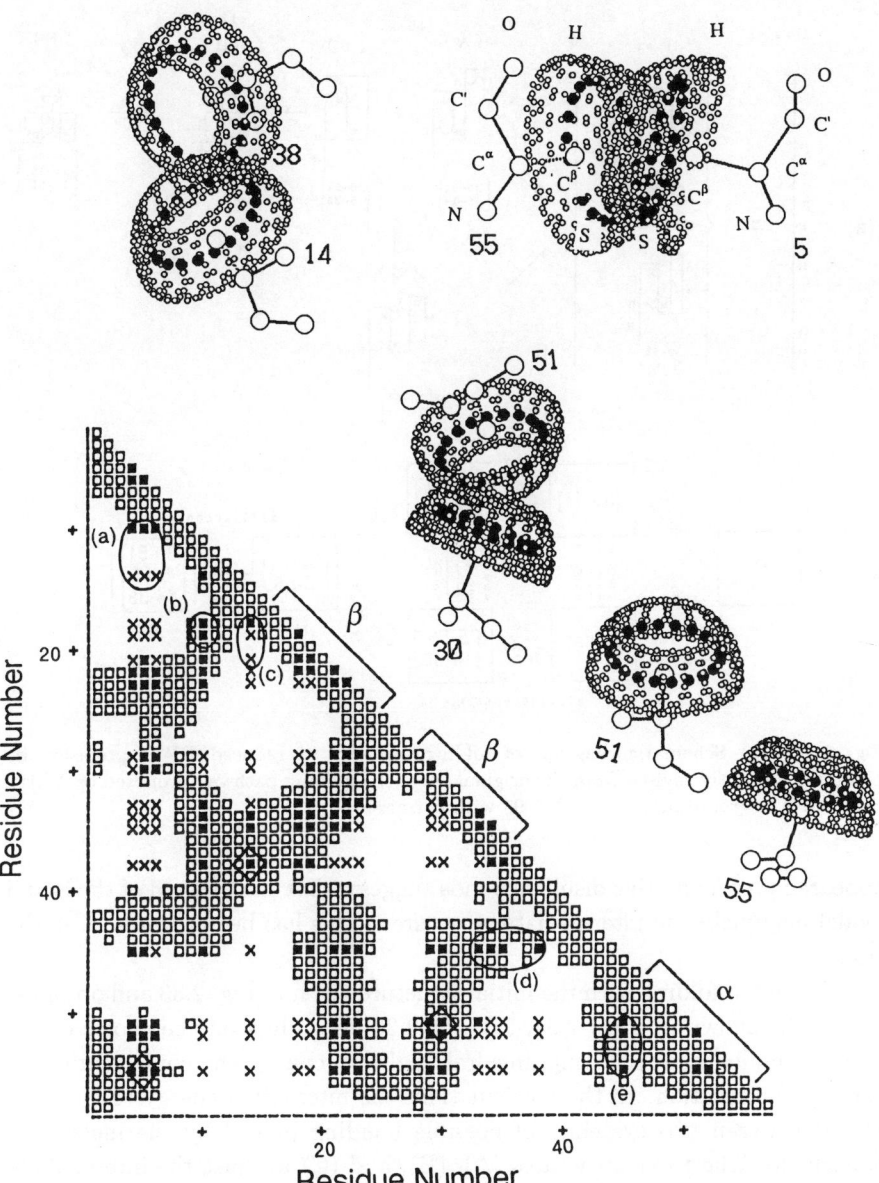

Fig. 2.31. Distance map of native BPTI (reproduced from Ref. 16 with permission).

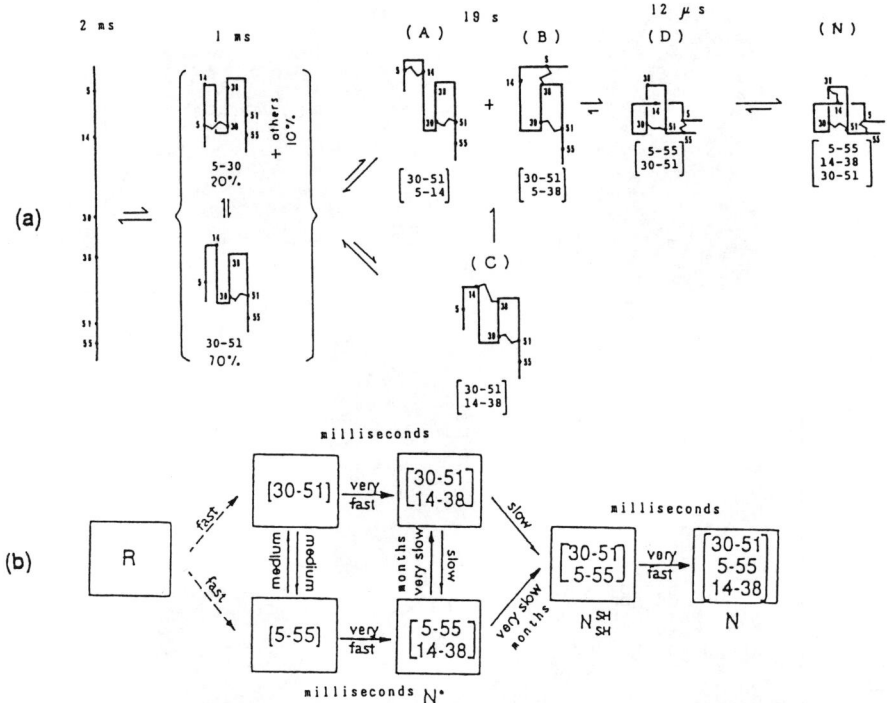

Fig. 2.32. (a) Schematic diagram of folding pathway of reduced BPTI proposed by Creighton, (slightly revised from the original one). (b) Folding pathway proposed by Weissman and Kim (reproduced from Ref. 95 with permission).

appearance of nonnative disulfide bonds suggests that the method of the island model may yield the intermediate structures. This has been proven to be the case.

We started folding with the initial structure given in Fig. 2.33 and obtained the results shown in Figs. 2.34, 2.35, or 2.36 depending on the choice of the random numbers for selecting dihedral angles to change the conformation by energy minimization. In these calculations the interaction energy (2.1) is employed between two cysteines of possible bonding judged by the method of lampshade. The three structures (A), (B), and (C) are just the intermediate structures identified by Creighton. To give more detailed explanations of the random choice of dihedral angles to vary the conformation in connection with the hydrophobic interactions, we notice the groups of hydrophobic residues

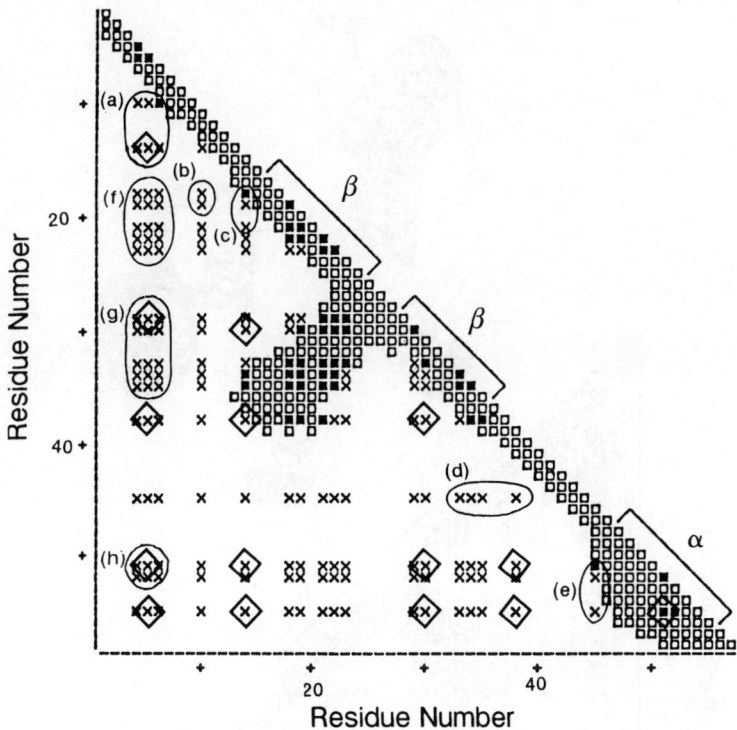

Fig. 2.33. Initial conformation for folding BPTI. Possible pairs of cysteines are indicated by (◇) (not necessarily to be bonded). Hydrophobic pairs (a), (b), ..., (h) are easily identified (reproduced from Ref. 16 with permission).

(a), (b), (c), ..., (h) in Fig. 2.33. The dihedral angles for binding groups (d) and (e) can be selected most easily and give rise to (30–51) disulfide bond, but not to the bond (38–51) or (38–55) due to their nonfacing lampshades. Next, consider the group (a). If the hydrophobic pairs in (a) are all bound, we have the structure (A) (Fig. 2.34), where the pairs in the group (f) are not bound. However, if a slight fluctuation of the conformation may loosen the contacts of the hydrophobic residues in (a) and fasten the contacts in the group (f), then we have the structure (B) (Fig. 2.35). The structure (C) is obtained by a small change in the orientation of Cys14, which causes the hydrophobic interaction in the groups (f) and (g) to loosen and to form the (14–38) disulfide bond by making Cys14 face to Cys38 (Fig. 2.36). Thus, the structures (A), (B), and

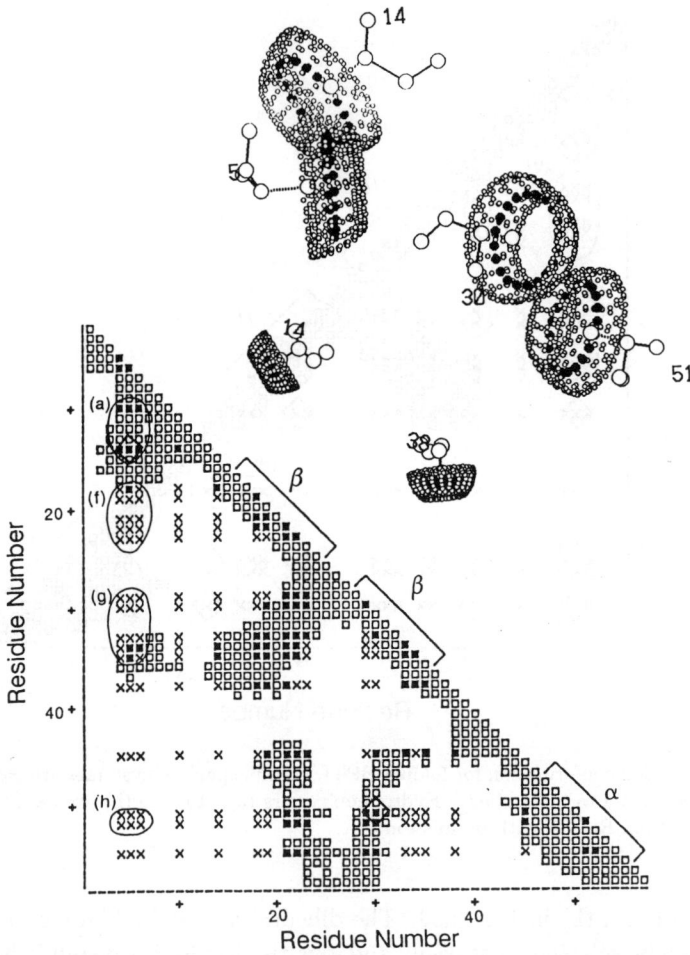

Fig. 2.34. Intermediate structure (A) of BPTI. Lampshades for (5–14), (30–51) and (14–38) are shown (reproduced from Ref. 16 with permission).

(C) are easily converted to each other and they are labile. Consequently, in the structure (A), when the disulfide bond (5–14) is broken and instead the disulfide bond (5–55) is formed, we have the structure (D) (Fig. 2.37). After that, the disulfide bond (14–38) can be easily obtained. A similar process will be found in the structure (B), where the disulfide bond (5–38) is broken and the disulfide bond (5–55) is formed to obtain the structure (D). The native

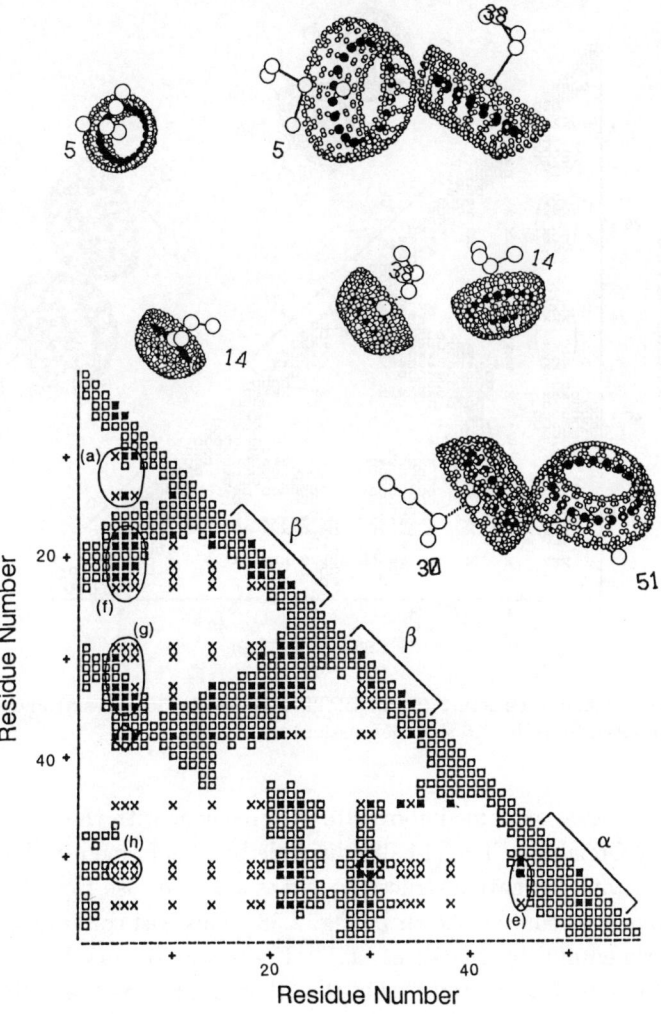

Fig. 2.35. Intermediate structure (B) of BPTI. Lampshades for (5–38), (30–51), (5–14) and (14–38) are shown (reproduced from Ref. 16 with permission).

disulfide bond (14–38) must not be made before the formation of (5–55), because an early formation of (14–38) bond causes the contact of hydrophobic residues in the group (i) (see the structures (B) and (C) given by Figs. 2.35 and 2.36 respectively) which prevents the formation of (5–55).

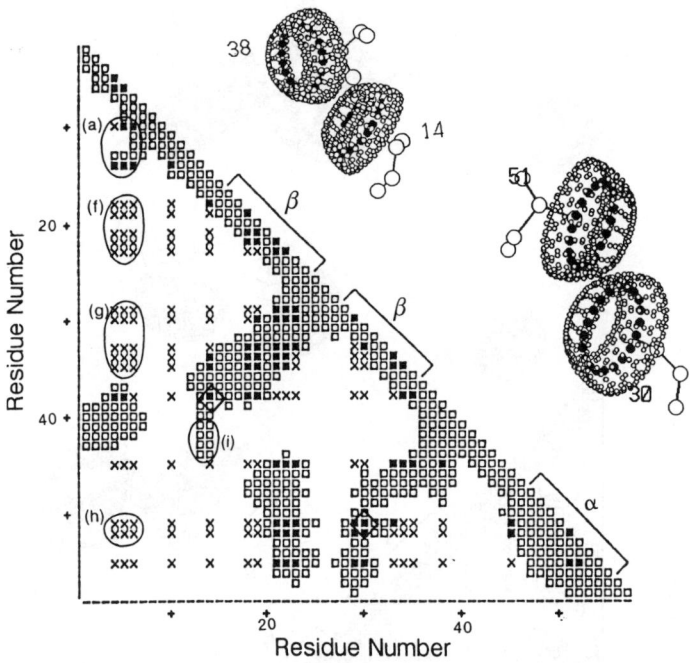

Fig. 2.36.   Intermediate structure (C) of BPTI. Lampshades for (14–38) and (30–51) are shown (reproduced from Ref. 16 with permission).

These processes explained above are in agreement with the folding process proposed by Creighton. The (14–38) disulfide bond is formed in the last stage via (D) to attain the native structure, and is also the case from (A), via (D) to the native structure as shown in Fig. 2.38. This was confirmed in various ways experimentally by Marks et al.,[101] Kress and Laskowsky,[102] Huber et al.,[103] and also by Darby et al.[104] On the other hand when the native BPTI is reduced, the intermediate (D) is obtained, indicating that the disulfide bond that is unbound first is (14–38).

Recently, however, a different folding pathway was proposed through new separation procedures of the intermediates, by Weissman and Kim[105,106] as shown in Fig. 2.32(b). This pathway does not involve the intermediates having nonnative disulfide bonds. However, they seem to have misinterpreted their experiments, causing them to reach wrong conclusions. This is because the intermediates (A) and (B) with non-native disulfide bonds are observed only

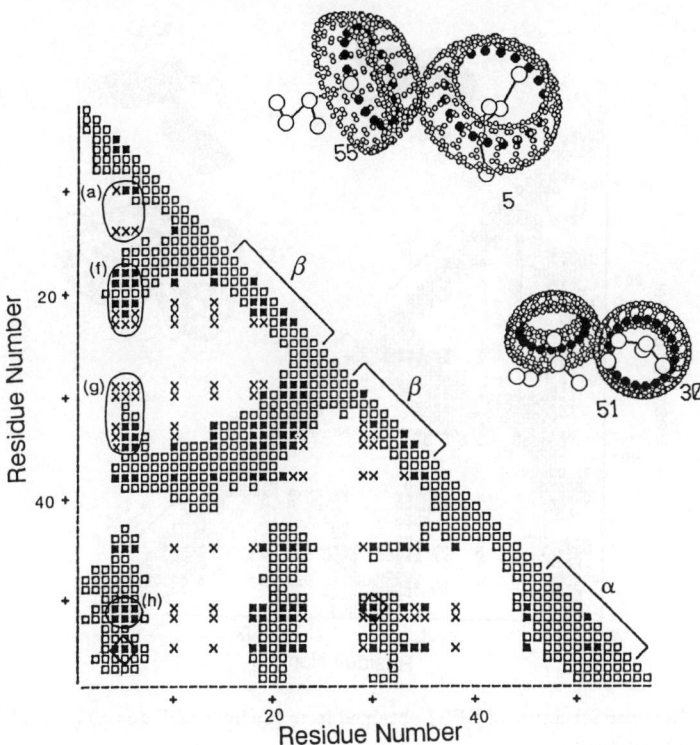

Fig. 2.37. Refolded structure (D) of BPTI. Lampshades for (30–51), (5–55) are shown (reproduced from Ref. 16 with permission).

for a few seconds in Creighton's experiments, but Weissman and Kim described the experimental results after a few minutes, during which the intermediates have already decayed. They also noticed the presence of (A) and (B) of very small quantities in their Fig. 5A in Ref. 106. Nonaccumulation of an intermediate does not imply the absence of the intermediates, because it simply indicates how rapidly the intermediate changes, as pointed out by Creighton.[107] Furthermore a clear explanation supporting Creighton's pathway is given by Goldenberg[108] from various experiments. Recently Creighton himself with his collaborators[109,110] have presented another evidence in favor of his original pathway, by engineering new species which have Ser in place of either Cys14 or Cys38. The kinetic studies of refolding these species, which are carried out with disulfide form of ditiothreitol (disulfide reagent designated as DTTS-S)

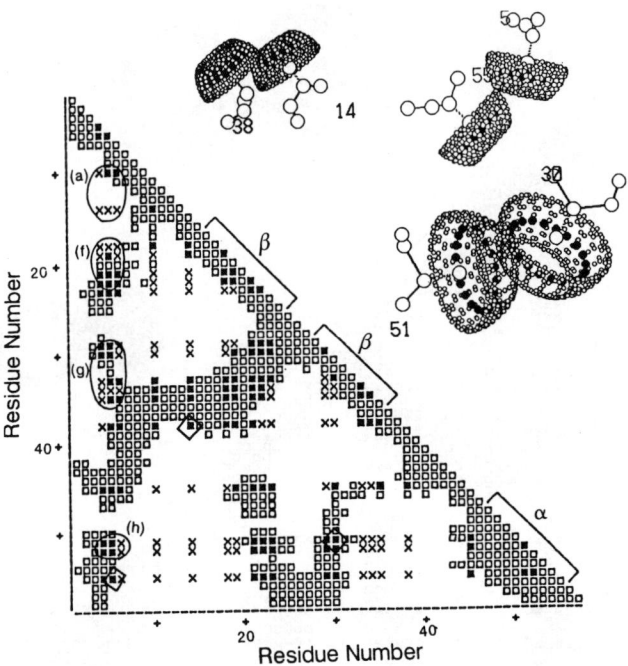

Fig. 2.38.   Refolded structure of BPTI obtained from the intermediate (A) (reproduced from Ref. 16 with permission).

reveals that two-disulfide intermediates similar to (A, Ser38), (B, Ser14) are easily obtained through one-sulfide (30–51) intermediate. They form the structure (D, Ser14 or Ser38) similar to (D) at an overall rate not much slower than that observed with the normal reduced protein, according to Creighton *et al.*. This implies that to reach the structure (D) from a one-disulfide (30–51) intermediate, the native disulfide intermediates, especially (30–51,5–14) and (5–55,14–38) are not necessary, contrary to Weissman and Kim's pathway. The experimental facts mentioned here in favor of Creighton's pathway are consistent with the theory presented here.

2.5   Flavodoxin and Thioredoxin

Flavodoxin and thioredoxin are examples of α/β proteins composed of α-helices and β-strands with β-α-β structure. The α-helix makes the two rod-like

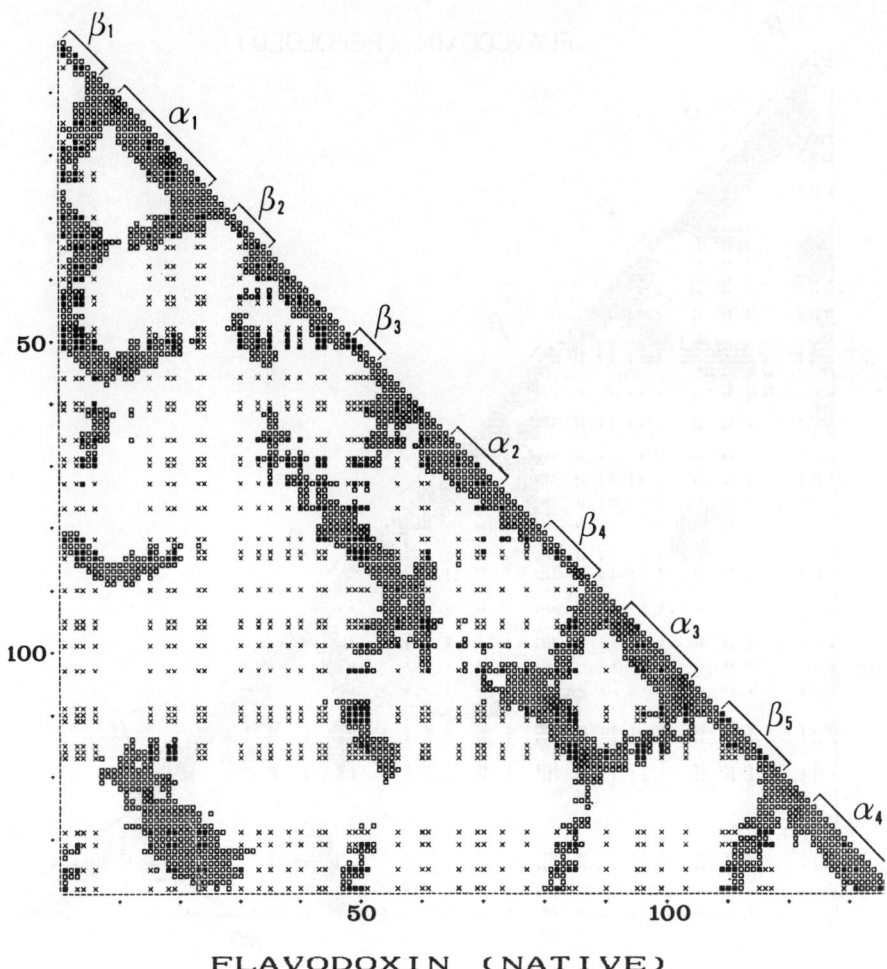

Fig. 2.39. Native structure of flavodoxin (reproduced from Ref. 13 with permission).

β-strands parallel each other. The interaction between the β-strand and the α-helix is none other than the hydrophobic interaction as can be checked in the distance map of the starting conformation having α-helices and β-strands (See Figs. 1.26 and 2.39).[13]

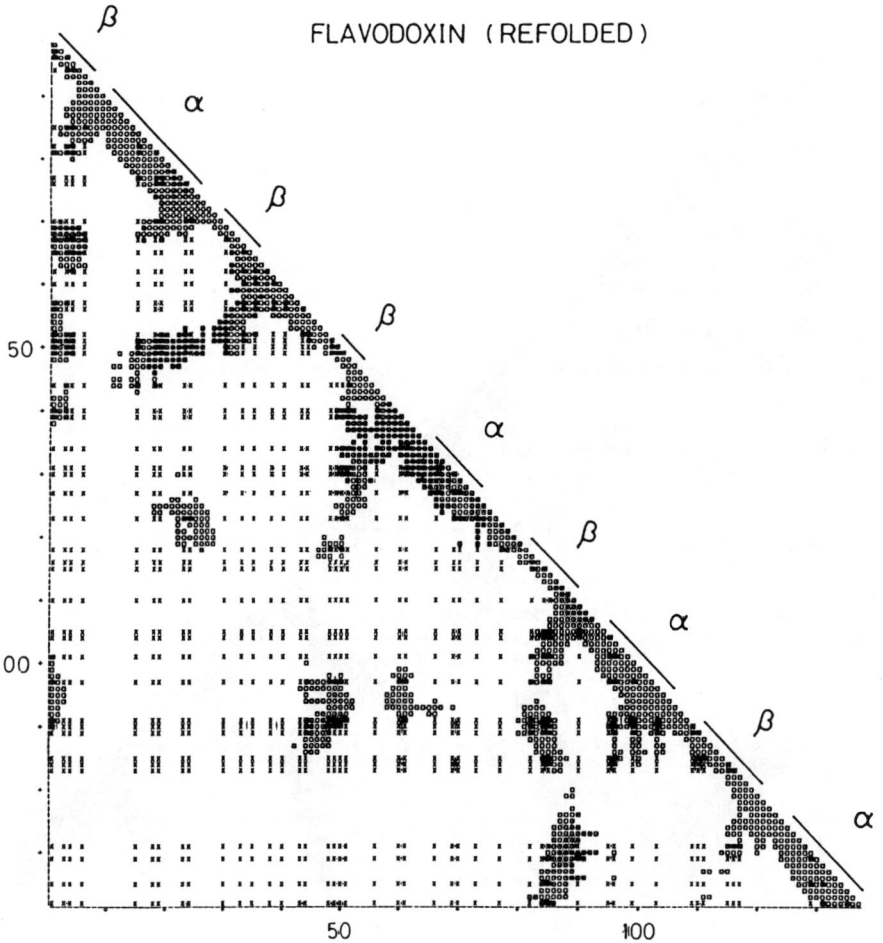

Fig. 2.40. Partially refolded conformation of flavodoxin (reproduced from Ref. 13 with permission).

Let us take the case of flavodoxin only, because thioredoxin can be refolded in the same way. Considering only the hydrophobic interaction responsible for bringing β-strands close to α-helices in the extended starting conformation of the α-helices and β-strands, we can obtain the conformation shown in Fig. 2.40. Although the pair β_1 and β_2, the pair β_3 and β_4, and the pair β_4 and β_5 are all brought close to each other, they are not parallel to each other.

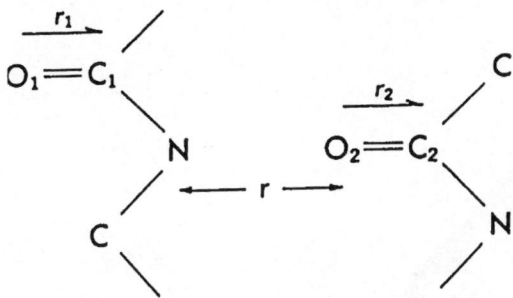

Fig. 2.41. Geometrical parameters in hydrogen bond in parallel β-structure.

Now we have to consider the hydrogen bonds between nearby β-strands. The hydrogen bond is usually short-range interaction, but the two β-strands, β_1 and β_2, in the conformation of Fig. 2.40 are almost 15 Å apart, so that the formation of hydrogen bonds is difficult, if the β-strands are assumed to be rigid and extended. However, in real proteins, the extended β-strands have some flexibility. Thus, the two β-strand are supposedly fluctuating, and thus may just happen to make contact at some point. If the hydrogen bond is then formed there, a zipper-like process can be expected with the result that the two β-strands make a parallel β-structure. Consequently, the hydrogen bond interaction appropriate for this situation can be modified to have long-range interactions with the following form (the units are kcal/mol for energy and Å for distance),

$$\phi(r) = \begin{cases} -3.0, & r \leq 2.9 \\ -3.0\exp\{-(r-2.9)/5.0], & 2.9 \leq r < 12 \end{cases} \qquad (2.2)$$

where r is the distance between N and O of the two strands, as illustrated in Fig. 2.41. When two β-strands make a hydrogen bond, as shown in Fig. 2.41, the two C=O bonds, O_1C_1 and O_2C_2 become parallel. Thus we denote the vectors $\overrightarrow{O_1C_1}$ and $\overrightarrow{O_2C_2}$ as r_1 and r_2 respectively, and multiply Eq. (2.2) by a factor

$$\theta = \frac{1}{2}\left(\frac{r_1 \cdot r_2}{|r_1||r_2|} + 1\right) \qquad (2.3)$$

to take directionality into account. This interaction is introduced for all pairs of N and O atoms between the two β-strands, which is supposed to result

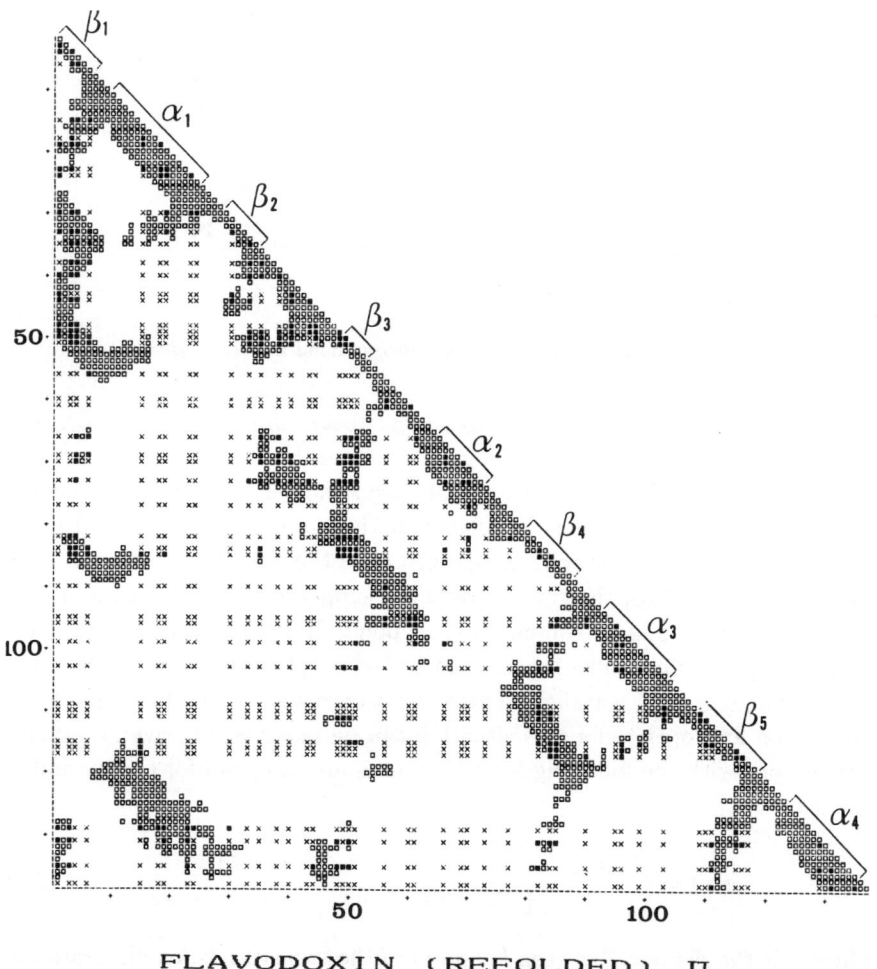

FLAVODOXIN (REFOLDED) Ⅱ

Fig. 2.42.   Refolded structure of flavodoxin (reproduced from Ref. 13 with permission).

in a parallel β-structure. The refolded structure, obtained by applying this procedure whenever two β-strands approach close to be parallel, is shown in Fig. 2.42, where one can notice that the two helices, α_1 and α_4 get closer than the case of Fig 2.40 to become almost parallel, as in the native structure (Fig. 2.39).

2.6 Ferredoxin

Ferredoxin is a protein which is structurally interesting from several aspects. It is composed of 54 amino acid residues including 8 cysteines, yet without disulfide bond. It has two iron-sulfur complexes, each of which contains 4 Fe, 4 inorganic S and 4 cysteine S atoms. According to the X-ray data (1FDX, 1976) no secondary structures are assigned (Recent PDB data (2000) shows, however, that ferredoxin from different source (new 1DUR replacing 1FDX) has 1 α-helix and 4 β-strands. But this does not alter or invalidate the following discussions). Our island model can be applied to any protein, and then the formation of small local structures is essential at the first stage of folding. It proceeds by hydrophobic interactions among hydrophobic amino acid pairs of the shortest distance along the chain (in other words, pairs capable to be bound most quickly) whether or not secondary structures exist. Thus we first tried to fold the fully extended structure as a starting conformation assuming all the dihedral angles flexible. But we could not obtain a structure similar to the native one, but obtained a compact one with DME of 6.84 Å. On the other hand, the DSSP program shows that α-helix, 3_{10}-helix and β-strands were assigned as shown in Table 2.1, contrary to the 1FDX data in PDB.

In fact by looking at the structure of the native ferredoxin shown in Figs. 2.43 and 2.44(a), (b) and (c), it was easy to see the presence of rather extended structures around the parts indicated as α-helix, 3_{10}-helix and β-strands by the DSSP program. Up to now we were able to refold many proteins with fairy good success, if exact secondary structures were known. Consequently we assumed segments around α-helix and 3_{10}-helix of DSSP as α-helices and those around β-strands as β-strands, and tried to refold the protein to see if we could obtain the structure similar to the native one. Thus our aim turns out to find the standard secondary structures which can give rise as correctly as possible to reproduce the native structure. After many trials of assigning secondary structure elements around the positions mentioned above, we

Table 2.1. The secondary structures obtained by the DSSP program.

AYVINDSCIA	CGACKPECPV	NIIQGSIYAI	DADSCIDCGS	CASVCPVGAP	NPED
CCEECTTCCC	CCTTGGGCTT	CCBCSSSCCB	CTTTCCCCCH	HHHHCSSCCE	ESCC

H: 4-helix, B: β-bridge, E: Strand, G: 3_{10}-helix, T: Turn, S: Bend, C: Coil.

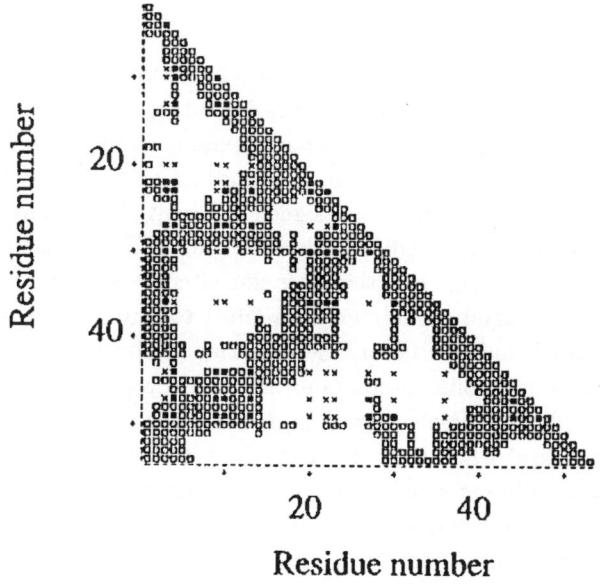

Fig. 2.43.   Distance map of the native conformation of ferredoxin.

reached the conformation shown in Fig. 2.45 of DME 6.34 Å, where the local structure, in other words, the short distance structure, was similar to the
native one, but the long distance structure was not good, as one sees in the
contact between the N and C termini. Consequently we next considered the
deformation of secondary structures by changing randomly the dihedral angles
of the residues in the secondary structures. A small change of one dihedral
angle may cause a big change in the protein conformation. Thus we searched
for the conformation of least energy so that hydrophobic residues in contact
remain bound and hydrophobic residues separated along a chain come close to
each other. The simulation method was simply random changes of standard
dihedral angles of the residues in the secondary structures. As a result, we
obtained Fig 2.46 with the DME of 4.29 Å, which shows a high prediction
accuracy. From this result, we can conclude that the secondary structures are
deformed at the last stage of the structure formation. In contrast to the target
proteins in our previous works,[11-16] the secondary structures of ferredoxin may
be so largely deformed from the standard conformations, that the PDB reports
absence of secondary structures in ferredoxin 1FDX. This means supposedly

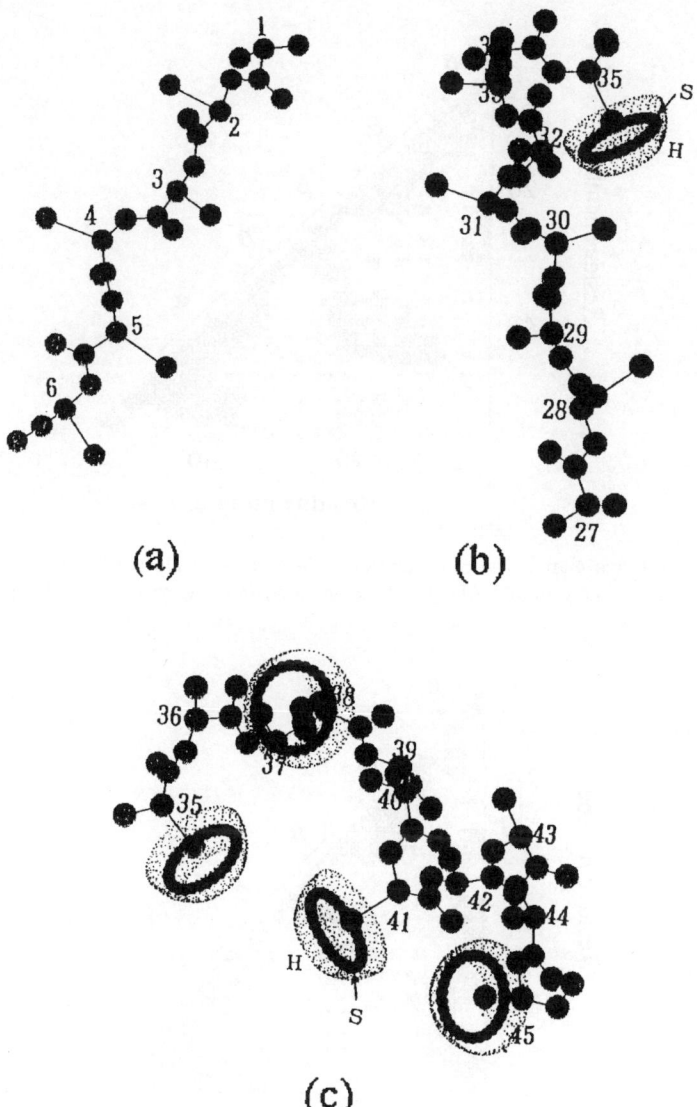

Fig. 2.44. (a) The structure of the 1st to the 6th residue region in the native structure. (b) The structure of the 27th to the 35th residue region in the native structure. (c) The structure of the 35th to the 45th residue region in the native structure. Cysteines are represented by "lampshade". The lampshades for bonded cysteines are face-to-face or crossed, while those of nonbonded cysteines are not, even if they are spatially proximate.

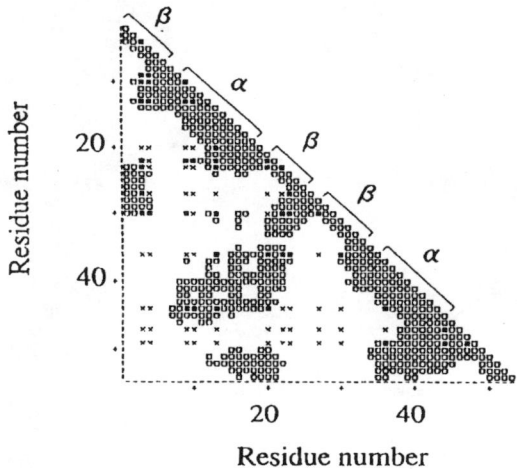

Fig. 2.45.   Distance map of the conformation of ferredoxin of DME of 6.34 Å. In addition to strong hydrophobic residues, Ala 10, 13, 48 are considered as weak hydrophobic residues.

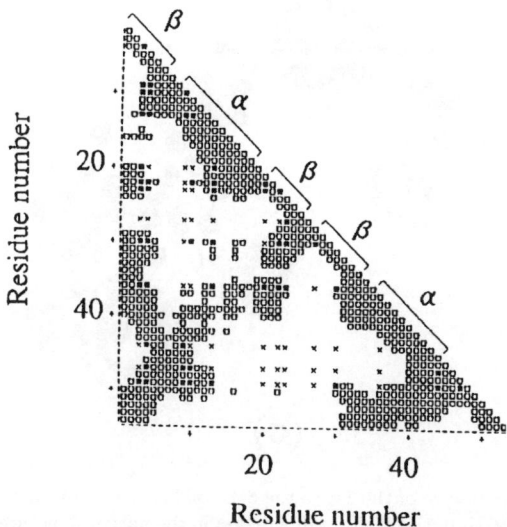

Fig. 2.46.   Distance map of the refolded conformation of ferredoxin of DME of 4.29 Å. The secondary structures were deformed by changing randomly the dihedral angles of the residues in secondary structures.

that the secondary structures are deformed largely in case of ferredoxin, by the interactions of the atoms surrounding the secondary structures.

The geometrical relation required for disulfide bonding is represented by the "lampshade".[15] Disulfide bonds are not formed in the native structure, because cysteines in the 35th to the 45th residue region cannot come close to each other (Fig. 2.44(c)). The same is true for the refolded structures simulated above.

Next we make mention of the iron-sulfur complexes in ferredoxin.[111] In the present work, however, we did not consider the presence of the iron-sulfur complexes. Thus our refolded structure is for apo-ferredoxin. The α-carbon backbone in the simulated conformation (Fig. 2.46) is shown in Fig. 2.47. One

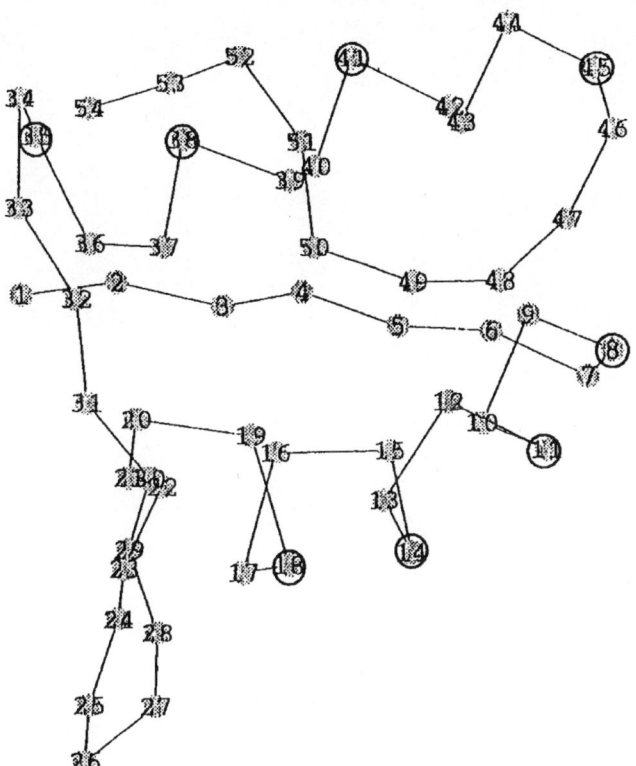

Fig. 2.47. The α-carbon backbone structure of the simulated conformation (Fig. 2.46). Notice a hollow around Cys18, 35, 38 and 41 and another one around Cys 8, 11, 14 and 45.

can see a hollow around Cys18, 35, 38 and 41 and another one around Cys 8, 11, 14 and 45. We conjecture that the two Fe_4S_4 ions can be bound to Cys35, 38, 18 and 41 and to Cys 8, 14, 11 and 45 respectively after the local structure formation around these cysteines at rather final stage of folding. When the iron-sulfur complexes are formed, the structure will become more compact, with the result of reducing DME value. Further in the native structure, Ile27 makes contact with many other residues (Fig. 2.44), but in refolded ones (Figs. 2.46 and 2.47), it does not. We suppose that Ile27 becomes in contact with other residues in the presence of the iron-sulfur complexes.

Chapter 3

Folding of a Protein of Unknown Structure

3.1 *Ab Initio* Method of Prediction of Protein Structure

3.1.1 *General principles*

We have so far discussed many details of mechanisms of protein folding to refold typical proteins. These details are necessary to predict the protein structure by the *ab initio* method. The method presented here has been obtained through these extensive studies and can be applied in principle to predicting the tertiary structure of any protein of unknown structure. We summarize the procedures in the following.

(i) *The determination of the secondary structures.* A statistical mechanical method for predicting α-helices and β-strands was presented in Sec. 1.2.4, where the determination of the weight parameters, 1620 for α-helices and 820 for β-strands, was necessary. Although our method can currently accomplish the secondary structure prediction with the average precision of about $68 \sim 70\%$, the results are not yet sufficiently reliable for packing the secondary structures into the correct tertiary structures. The most difficult part in predicting protein structure lies now in this rather poor prediction of the secondary structures and the antiparallel β-structures (see Appendix D). Therefore, at present, whenever we want to fold, we start to employ any method, for practical purposes, to determine the secondary structures, if available. In Sec. 3.1.3 an example will be presented.

(ii) *Hydrophobic pairs.* As we discussed in Sec. 1.3, the driving force of performing the packing of the secondary structures is the long-range hydrophobic interaction. To see its role, first we construct the distance map of the initial

conformation that has the secondary structures of standard molecular parameters, such as α-helices, β-strands, and β-sheets. The other parts of the chain
are made to have extended conformation. The amino acid residues Leu, Ile,
Val, Met, Trp, and Phe are considered to be hydrophobic, and the same symbols □, ■, and × as in Figs. 1.15, 1.25, 1.26, etc. are used. When necessary, in
particular when hydrophobic residues are scarce, weak hydrophobic residues
Ala have to be included. Then looking at the distance map of the initial conformation, we pay particular attention to the distribution of hydrophobic pairs
marked as ×. Among them, those essential for packing of nearby secondary
structures are located close to the diagonal. More precisely, the residues of
hydrophobic pairs located close to the diagonal are at a short distance and
thus are bound easier than the pairs at a longer distance. We circle these
hydrophobic pairs marked as × as we did in Figs. 2.2, 2.7, 2.20, and so on.

(iii) *Coarse grained model of amino acid residues.* Simulation of protein
folding proceeds by searching for the conformation of lower energy in a restricted space, as discussed in the preceding sections. At the earlier stage of
folding we do not have to estimate the energy precisely, but instead we have
only to calculate it with a simplified model. To do this, we replace the side
chains of amino acid residues, other than glycine and alanine, by spheres of appropriate radii and locate them at the appropriate positions. The geometrical
parameters are listed in Table 1.6, which are the averages of nine proteins. The
values are usually different from those of a particular amino acid residue of a
protein, but this does not seriously affect the resulting conformation. When
necessary, they can be changed to more realistic side chains at the last stage
of folding using QUANTA/CHARMm or PRESTO (see (vii)).

(iv) *Packing order of the secondary structures.* The packing of the secondary structures is performed by seeking the state of the lowest energy by
searching randomly the space spanned by the relevant dihedral angles. The
procedures are described in Appendix C. This state is realized by binding
the hydrophobic residues. Consequently, the most quickly bound hydrophobic
pairs are those that the chain connecting the pairs has least number of free
amino acid residues. By the word "free" we mean that the amino acid residue
does not lie in the secondary structures. This yields the order of packing of
secondary structures. What happens when the packing order is violated will
be discussed in Sec. 3.1.2.

(v) *Energy calculation.* To calculate the conformational energies of globular proteins in water, we take the following into account: the Lennard-Jones

interactions between nonbonded atoms, hydrophobic interactions between hydrophobic residues, electrostatic interactions between charged residues, if any, or between polar residues, hydrogen bonding energy, and torsional potential around chemical bonds, with the bond lengths and angles kept constant. Among them, the first two are most essential, i.e., for avoiding collapse of molecules, the Lennard-Jones potential, whose parameters were determined by Momany *et al.*[84] and for the driving force of packing, the hydrophobic potential in the form of Eq. (1.20). The hydrogen bonds are important, especially for the formation of secondary structures. Electrostatic interactions are sometimes reduced in water, because of a large dielectric constant of water. But the interactions other than the first two mentioned above can usually be ignored in the calculation and can be considered at the final stage (see (vii)).

(vi) *Disulfide bonding.* Even when two cysteine residues come close together, a disulfide bond cannot necessarily be formed if appropriate mutual orientation is not satisfied. This criterion can easily be verified using lampshades. The lampshade criterion was discussed in Secs. 2.3 and 2.4. We introduce the disulfide bonding potential in the form of Eq. (2.1) between the relevant S atoms every time two nearby cysteines are found to satisfy the above criterion.

(vii) *Refinement of the structure.* We can proceed with the packing of the secondary structures by further seeking the conformation of least energy through the smaller and smaller changes in the dihedral angles of free residues by means of Bremermann's method (Appendix C). In doing this, we do not have to worry about the possible presence of deep local minima, which are sometimes considered to be the difficulty in the *ab initio* method of prediction. This is because the effective conformation space turns out to be a small part of the whole conformation space restricted by the hydrophobic interaction. Anfinsen's dogma holds in this smaller space, and the Levinthal paradox can be avoided, while the deep local minimum is expected to be absent (see Sec. 1.3.7). Otherwise a polypeptide with such an amino acid sequence would not be a protein of biological significance. At this stage of the folding process, we introduce nativelike side chains instead of the spheres at the positions of the residues and revise the conformation by slightly modifying the molecular coodinates, especially by deforming secondary structures to reach the much lower energy state. The software QUANTA/CHARMm or PRESTO is useful for this purpose.

Table 3.1.   Comparison of the two conformations obtained from different packing order in cytochrome b_{562}.

Packing order of the pair of α_2-helix and α_3-helix among other pairs	Energy (kcal/mol)	DME (Å)
Last	−1044.25	3.28
First	−1132.03	4.55

3.1.2   *Packing order and Anfinsen's dogma*

In Sec. 3.1.1(iv), the importance of the packing order is pointed out. Now, we will show some examples of the conformations obtained by simulating a wrong folding order.[114] In Sec. 2.2.2 cytochrome b_{562} was refolded by the method presented here. The packing between α_2-helix and α_3-helix was last in the folding order. Instead, if one tries to pack the α_2- and α_3- helices first, a wrong conformation different from the native one is obtained, but the conformation energy is less than the native structure, as shown in Table 3.1.

The same occurs in the N-terminal domain of phage 434 repressor of 63 amino acid residues.[115] The starting conformation for packing is shown in Fig. 3.1, where 5 helices (A, B,..., E), nonhelical coil parts (1), (2),..., (6), and the important hydrophobic pairs (a), (b), (c), and (d) are indicated. We are interested in the folding order of the coil parts (2),..., (5) and their hydrophobic residues for driving packing (a), (b),..., (d), since the coil parts (1) and (6) are irrelevant. Examining Fig. 3.1 in detail, we obtain Table 3.2 for the hydrophobic pairs, their pertinent amino acid residues responsible for changing the spatial distances of the pairs, and the number of the dihedral angles (number of freedom) for each pair. One sees that the folding order should be (2), (3) and then (4), (5) shown in the last column of Table 3.2.

Ignoring for a while the above consideration, we tried to fold phage 434 using various folding orders with the results shown in Table 3.3. Since the packing of A and B and that of D and E are almost independent of the packing of other parts, the important packing order is the coil part (3) followed by the coil part (4). This is really the case, as can be seen in Table 3.3; that is, the trials 1, 2, and 3 give good DME values while the trial 4 gives rather big DME,

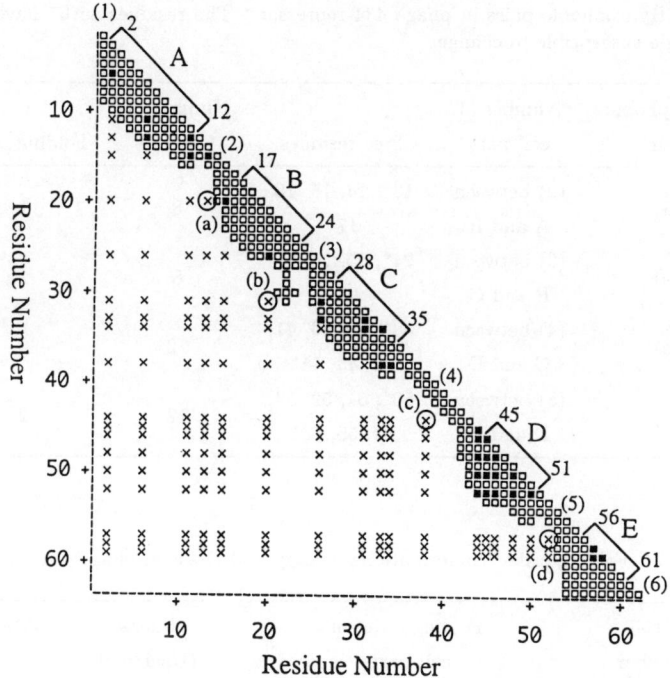

Fig. 3.1. Starting conformation of phage 434 repressor.

but lowest energy. The most preferable folding order of the coil parts will be $(2) \to (3) \to (4) \to (5) \to (1) \to (6)$, but this is not yet confirmed.

Another case study of folding order is sea hare myoglobin (see Sec. 2.2.3). It was shown that in this case also, one of the wrong folding orders gave lower energy than, and, notwithstanding, almost the same value of DME with, the right one (for detail see Ref. 20).

Folding through wrong packing order is realized by computer simulation, but not in real experiments because of the restriction of conformation space imposed by hydrophobic interactions realized under usual experimental conditions. In this restricted region, Anfinsen's dogma is valid. Out of this region, a lower energy state may or may not be realized. The former case is cytochrome b_{562}, phage 434 repressor and sea hare myoglobin, but the latter case has not yet been found in our not so exhaustive computer experiments. In the next section, an example of protein structure prediction will be presented.

Table 3.2.   Hydrophobic pairs in phage 434 repressor.* The residues with* have only one dihedral angle susceptible to change.

Hydrophobic pair	Number of coil part	Free residues	Number of freedom	Folding order
(a)	(2) between A and B	13*, 14, 15, 16, 17*	8	1
(b)	(3) between B and C	24*, 25, 26, 27, 28*	8	1
(c)	(4) between C and D	38*, 39, 40, 41, 42, 43, 44*	12	2
(d)	(5) between D and E	50*, 51, 52, 53, 54, 55, 56*	12	2

Table 3.3.   Folding order, conformational energy and DME in phage 434 repressor.

Trial number	Folding order of coil parts	Energy (Kcal/mol)	DME (Å)
1	$(1) \rightarrow (3) \rightarrow (5) \rightarrow (6) \rightarrow (4) \rightarrow (2)$	−524.1	3.91
2	$(3) \rightarrow (2) \rightarrow (1) \rightarrow (5) \rightarrow (6) \rightarrow (4)$	−445.2	4.14
3	$(1) \rightarrow (3) \rightarrow (4) \rightarrow (2) \rightarrow (6) \rightarrow (5)$	−496.9	4.10
4	$(2) \rightarrow (1) \rightarrow (4) \rightarrow (3) \rightarrow (6) \rightarrow (5)$	−527.7	5.45

3.1.3   *Application to parathyroid-hormone-related protein (residues 1–34), abbreviated as PTHrP(1–34)*

As an example, we take PTHrP(1–34),[19] which has the primary structure as shown in Fig. 3.2.

The process of folding this small protein will be described fairly in detail to demonstrate how to use the method presented in the above sections. We assume that, as mentioned at the end of Sec. 3.1.1.(i), the 3rd through the 9th residues form an α-helix as is assigned by Barden and Kemp[115] by means of NMR experiments. Figure 3.3(a) is the distance map of the conformation of PTHrP(1–34), where the 3rd through the 9th residues are in an α-helix and

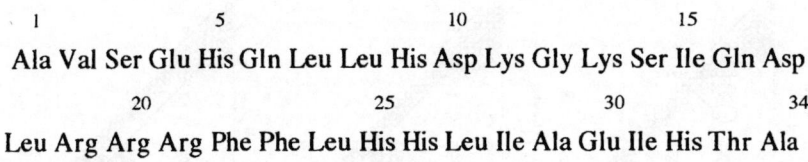

Fig. 3.2.   Amino acid sequence of PTHrP(1-34).

others are in the extended conformation. We consider the residues Leu, Ile, Val, Met, Trp, and Phe to be hydrophobic. Their pairs at short distances are circled in Fig. 3.3(a). Among them, we first consider the pairs (Leu18–Phe22), (Phe23–Leu27), (Leu24–Ile28), and (Leu27–Ile31) of 4-residue distance in the hydrophobic groups (b), (c), and (d) as the pairs of the shortest distance and introduce the hydrophobic interaction of the form of Eq. (1.20). According to the general principle (iv) mentioned in Sec. 3.1.1 these pairs will be bound at the earliest stage of folding. Thus, the dihedral angles of the residues in Arg19–Arg21 and Leu24–Glu30 are considered, because they are the variables necessary and sufficient for binding the hydrophobic pairs and leading to the lowest energy state. Energy calculations are performed through hydrophobic, as well as 6–12 Lennard-Jones interactions every time certain values are assigned to the dihedral angles by generating a set of random numbers. A search for the state of the lowest energy is made as described in Appendix C.

After about 150 steps of this minimization, the calculated energies become almost unchanged, but we continued up to 400 steps to confirm this. From the same initial conformation, the process of minimization was carried out 30 times, using different random numbers. After these calculations, we selected three possible structures at this intermediate stage by referring to total energy, hydrophobic energy, and conditions of hydrophobic bonding. One of them is shown in Fig. 3.3(b). In this structure, the pairs of hydrophobic residues under consideration are bound as expected. Next, we tried to make (Leu18, Leu24) of the group (b), (Phe22, Leu27), and (Phe23, Ile28) of the group (c) contact, setting interaction distances within 5 residues and changing the dihedral angles of Arg19–Leu27. Calculations were carried out from each selected structure, continued up to 800 steps, and tried 30 times, using different random numbers. After these calculations, we selected a few structures from 90 structures by means of the criteria mentioned above. Figure 3.3(c) is one of them where the hydrophobic pairs of our present concern were bound. The same procedures

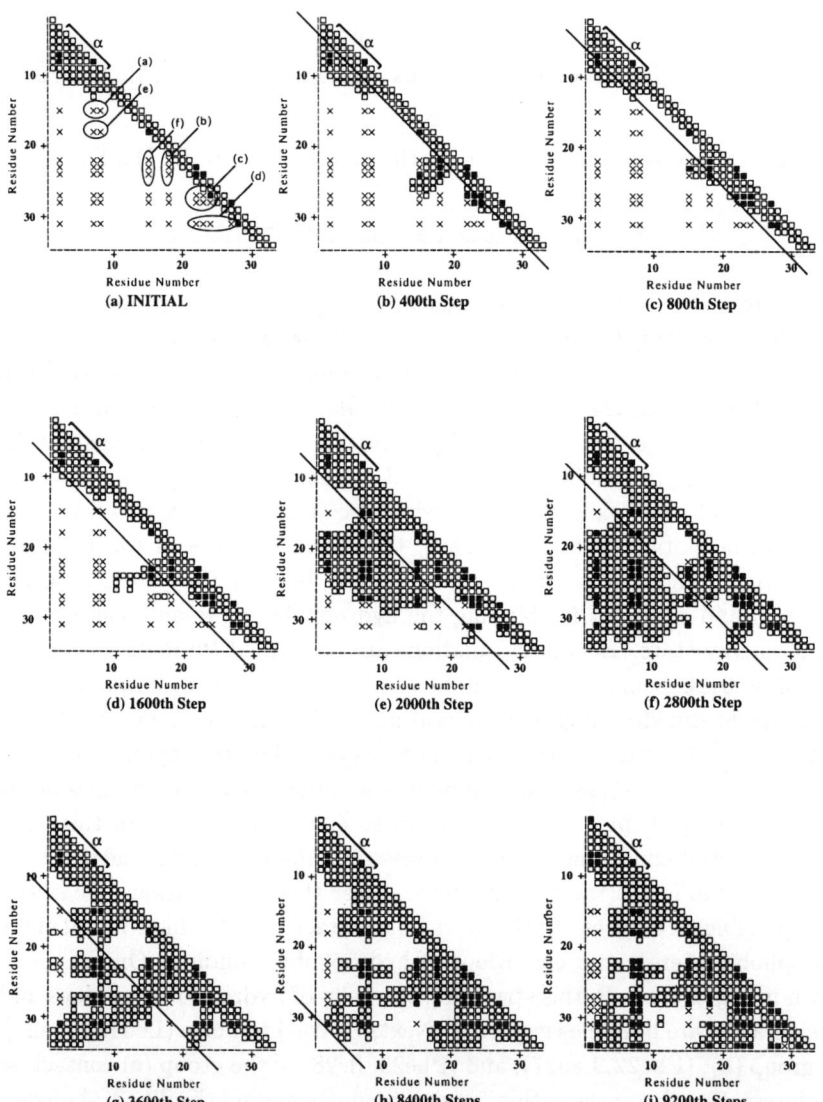

Fig. 3.3.   Distance map of PTHrP(1-34). (a) Starting conformation. (b) $\sim$ (i) Intermediate conformations at the various folding steps. The pairs lying in the region between the diagonal and a line parallel to it have distances less than the value determined by the parallel line. The interaction energies are considered at each step between the residues of pairs (reproduced from Ref. 19 with permission).

were iterated up to 8400 steps. Folding proceeded from Fig. 3.3(d) to 3.3(h). In Fig. 3.3(e), α-helix was in contact with the middle part of the chain by the hydrophobic interactions between Leu7, 8 and Ile15. The hydrophobic interaction between Leu8 and Leu18 built up the global frame of the structure (see Fig. 3.3(f)). The minimum energy calculations were carried out by gradually

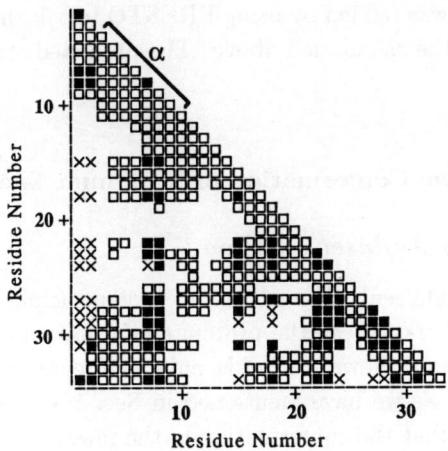

Fig. 3.4. Distance map of the predicted structure of PTHrP(1-34) (reproduced from Ref. 19 with permission).

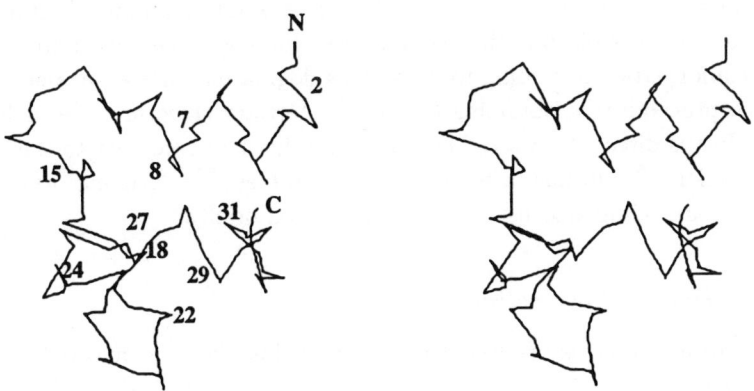

Fig. 3.5. Main chain of the predicted structure of PTHrP(1-34) (reproduced from Ref. 19 with permission).

changing the interaction range, indicated by a line parallel to the diagonal, to finally reach Fig. 3.3(h). At this stage (8400 steps), we had no more hydrophobic residues to bring N- and C-termini close together. Consequently, Ala1, 29 and 34, which are regarded as weak hydrophobic residues, were taken into consideration for energy calculation up to 9200 steps.

The best structure among those thus obtained is shown in Fig. 3.3(i). Finally, the structure was refined by using PRESTO to take into account various energies ignored in the calculation above. The predicted structure is shown in Figs. 3.4 and 3.5.

3.2   Search for the Conformation of Minimum Energy

3.2.1   *Validity of Anfinsen's dogma*

Establishment of Anfinsen's dogma, or the first principle of protein folding, urged many researchers to seek the protein conformation of least free energy. This Anfinsen's dogma, however, holds only in the restricted region of the conformation space, as we have mentioned in Sec. 1.3.1. We have also mentioned in Sec. 3.1.2 that the conformation of the lowest energy over the whole conformation space is not necessarily the native conformation of the protein, as verified in the cases of cytochrome b_{562},[18] phage 434 repressor[114] and sea hare myoglobin.[20] Furthermore, the landscape of energy profile in conformation space is supposed to be very complicated, having deep minima and steep hills, which make it difficult to search for the absolute minimum, resulting in the violation of the second principle of protein folding. Thus, the simple search of the conformation according to Anfinsen's dogma may not always be the legitimate procedure for obtaining the correct tertiary structure. Nevertheless, much effort is directed towards this sort of study. Thus, we refer to the recent review article[116] and new attempts at this problem.[50,51] Among others, the simulated annealing algorithm will be mentioned briefly.

3.2.2   *Simulated annealing*

Optimization problems are found in various fields closely connected to computer science and technology. Usually, the problem is reduced to finding the least value of a suitably chosen objective function. In this respect, it is similar to the problems of statistical mechanics where the equilibrium state is the lowest free energy. A computer algorithm for this purpose was first developed by Metropolis *et al.*[118] in statistical mechanics and then this idea was transferred to more general multivariate optimization problems and a new technique of simulated annealing developed there was returned to statistical mechanics. Although this technique is rather well known, a brief exposition of the simulated annealing will be appropriate for some readers, and thus, will be given here in terms of statistical mechanics.

A search for the lower energy state can be done through employing thermal fluctuations in the system by means of the Monte Carlo simulation and letting the system wander over the various states. However, when this process is extended over the conformation space of complicated energy profile, it will be difficult because of the presence of many local minima where the system will easily be trapped. Thus, a simple application of Metropolis algorithm at a definite temperature is not effective in the optimization problems and an improved procedure is required. Crossing over the high barriers around a local minimum of energy is made possible when the temperature is raised and the system is kept at this temperature for a while (annealed). Then, the system takes the lower energy, on the average, at the thermal equilibrium. On decreasing the temperature gradually, the system takes lower and lower energy states because it cannot cross over the barriers to the high energy state. This process can be continued by decreasing further the temperature and finally, it is supposed to reach the lowest energy state. Here lies the origin of the method of simulated annealing.[119] This method was applied to the spin glass[120] as an example of solid state physics and then to protein conformation[121-124] in accordance with Anfinsen's dogma. Further, another method called the multicanonical algorithm, which was developed by Berg and Neuhaus,[125,126] was applied to proteins by Hansmann and Okamoto.[127,128] These methods were successfully applied to small proteins such as Met-enkephalin. But they made no mention of the role of the hydrophobic interaction that had been attributed as the key factor in protein folding from various experiments as discussed in Sec. 1.3.2. Thus detailed mechanism of protein folding is not unveiled. Most importantly, Anfinsen's dogma holds in a restricted space governed by hydrophobic interactions (Sec. 3.2.1).

Chapter 4

Topics Related to Protein Structures

4.1 Phase Transition

Phase transition is an important subject of statistical mechanics. As explained in Sec. 1.2.2, it is characterized as a singular point of certain thermodynamical quantities in the thermodynamic limit, i.e. in the limit of infinitely large value of N (number of particles) and V (volume or system size) keeping N/V constant. In a finite system singularity does not take place but a diffuse phase transition can be expected. A diffuse phase transition is also observed in one dimensional systems as already discussed, if it has cooperativity in the interaction. The tertiary structure of a protein is brought about by a diffuse phase transition. Phase transition, as well as diffuse phase transition, usually gives rise to new properties in materials science such as ferromagnetism, superconductivity, etc., and new functions in biological systems such as enzyme activity, material transport of membrane, etc. Furthermore, the new state obtained by the result of a phase transition is usually stable; ferromagnetism is kept unchanged up to the Curie point and enzyme is active over a rather wide range of the environmental conditions without significant changes in its activity, because otherwise the life cannot survive. Another aspect of the phase transition is the ability of regulation, as one known in the case of water and ice systems to keep the temperature at $0°$C. This system undergoes a first order phase transition and has heat of melting enabling by itself the role of regulation. In a second order transition, or in a diffuse transition, heat of transition or something else cannot be expected, but the existence of a fairly definite phase transition point is sufficient to prepare a device of regulation. Supposedly, this sort of device may be found in biological systems, for example in homeostasis, as a kind of protein function.

4.2 Module

4.2.1 *Exons, introns, and modules*

In 1977, Berget, Moor, and Sharp[129] found that the genes in eukaryotes are split into exons and introns. The DNA base sequence is first copied into the messenger RNA, where the introns are removed, and exons are spliced together. The role of exons and introns in evolution was discussed by Gilbert[130] as performing shuffling of exons in DNA, which promotes the evolution of proteins. This implies that exons are considered to supply the units of protein structure, which are rather stable and independent. In fact, Gō[80] discovered the constructing units, or modules, of protein, which are separated each other inside the tertiary structure. This was achieved by studying distance map that described the pairs at long distances more than 25Å apart, contrary to those at short distances employed to investigate the folding interaction. She found four modules: A_1, A_2, A_3, and A_4, which were separated and supposed to behave independently, in β-chain of mouse hemoglobin having three exons. A_1 and A_4 were coincident with first and third exons respectively. Consequently she conjectured that one module should correspond to one gene, and suggested that an intron between the exons corresponding to A_2 and A_3 had been missing during evolution and thus A_2 and A_3 had their own exons, respectively. This was really found to be the case, because the intron was discovered in leghemoglobin.[131] Since then, the correspondence between exons and modules have been confirmed in various proteins.[132] The method of centripetal profile was also developed for identifying modules from the crystal data of the structure of a rather large protein.[133] Introns were first found in eukaryotic genes and it was supposed that procaryotic genes were not split by introns. Recently, however, introns have been found in some bacterial genes,[134,135] implying that they were lost in most contemporary procaryotic genes and also suggesting an interesting problem of molecular evolution.[136]

4.2.2 *Stability of modules*

Modules must be stable, considering their roles in protein evolution. This requirement is satisfied, because the modules in a protein are independent of each other since they are separated more than 20Å in the protein. Their conformations must be unaffected by other modules. In other words, when the modules are excised from the protein, their conformations are supposed to be kept almost unchanged. This was experimentally verified by Ikura *et al.*[137] for barnase from *Bacillus amyloliquefaciens*, which has 6 modules. They chemically synthesized the three peptides 1–24, 24–52, 52–73 corresponding to the modules with one residue overlapping with the neighboring module. The solution NMR studies were carried out, with the result indicating that the conformations of the isolated modules are almost the same as those observed in the crystal structure. The differences are supposed to have arisen from the smallness of the molecules, with possible fluctuation. Thus, we can expect the conformation of the module to obey the same folding mechanism of ordinary proteins.

The stability of modules comes from various interactions in the module, among which hydrophobic interaction is essential because this interaction gives rise to the stable structures of modules as parts of a protein as was emphasized above. However, since many of the hydrogen bonds are located within a module, Noguti *et al.*[138] consider that the intra-module hydrogen bonds may contribute to the stabilization of the module. This situation is to be compared with the stability of α-helix, where the medium-distance hydrogen-bonds are decisive for its stability as discussed in Sec. 1.2.6, although the contribution of nonbonded interactions are rather predominant. The role of hydrogen bonds in the α-helix is replaced, in the case of modules and proteins, by the hydrophobic interaction at medium- or long-distance. The short-range hydrogen bonds are formed in a module with the help of the long-range hydrophobic interaction.

4.3 Molecular Chaperones

Anfinsen's dogma and subsequent research explained in detail in the above sections are based on numerous experiments performed *in vitro* especially in dilute systems. It was shown that the folding of a nascent polypeptide into an active protein proceeds spontaneously, without any help of enzymes. Until recently, this conclusion was believed, more or less implicitly, to hold *in vivo* or in the cytoplasmic systems. The discovery of the presence of intervening proteins called molecular chaperones, such as heat shock proteins (HSP) in *in vivo* folding, was a heart-shocking event to the researchers standing on thermodynamical side. The system *in vivo*, however, is usually not dilute, but is complex, containing many ingredients. Thus, how the protein folding proceeds *in vivo* and whether or not the validity of Anfinsen's dogma still hold there, are the subjects to be investigated. Various experiments on protein folding *in vivo* revealed that it required some other proteins (molecular chaperone coined by Laskey *et al.*[139]) and ATP. Furthermore, molecular chaperones have been found to play several functions, such as transporting protein molecules through membrane, regulating the activities of transcription factors of stress proteins, conformation change of prions, etc. We will here discuss the role in protein folding only.

Nascent proteins or denatured proteins have hydrophobic residues exposed to the environment and thus they are easy to aggregate *in vivo* in a concentrated state, and unable to fold by themselves into the native structure. Molecular chaperones are supposed to prevent aggregation by trapping the unfolded proteins or the molten globule state proteins, where the secondary structures are easily formed immediately, through the unbound hydrophobic residues of flexible unfolded parts, separating each protein from the others. The isolated polypeptide chains are thus able to fold into the native state by the process described in the above sections. Consequently, Anfinsen's dogma observed *in vitro* experiments remains valid. The purpose of this section is to emphasize the usefulness of the thermodynamical approach and a brief mention will be made on the mechanism of molecular chaperones in connection with their structures.

Among the chaperones, the most well-investigated one is chaperonin of *Echerichia coli* (GroEL/GroES), a complex of two kinds of HSP, GroEL and GroES. The three-dimensional structure was determined by X-ray analysis and by electron microscope.[140−142] GroEL and GroES each form layers of seven subunits arranged side by side in the form of a ring, but the former is bigger and has molecular weight of 58kD while the latter has 10kD. The main structure of chaperonin is composed of two layers of GroEL one of which has two ADP and GroES. Therefore the structure has a cavity inside and the layer of GroES is just making a cover for this hole. The mechanism of chaperonin-mediated protein folding was proposed by Mayhew *et al.* and by Weissman *et al.*[143−146] The protein in molten globule state is adsorbed primarily by long-range hydrophobic interaction at the periphery of the layer of GroEL facing the cavity and in *trans* of GroES. The GroES is removed from the ring of GroEL, but rebinds together with ATP to the ring of GroEL at the other side in *cis*.

It is supposed that the hydrolysis of ATP yields energy to loosen the binding of the polypeptide and the chaperonin structure itself, resulting in free motion of the polypeptide to fold spontaneously into the committed structure in the widened cavity or outside of it. The polypeptide confined in the cavity will suffer more or less from the restriction of its free motion, similar to the case of the formation of the loop structure in membrane proteins as will be discussed in the next section. If the chain has sufficient flexibility, this restriction of free motion does not cause any difficulty for correct folding, since the early steps of the folding of the polypeptide are to bind closer hydrophobic pairs. Otherwise, the folding will take place outside the cavity after the polypeptide is ejected due to the partial loosening of the chaperonin structure. ADP is also thought to play some role in stabilizing the structure, but the more details of the whole scenario still remains unknown. Nevertheless, we can believe the validity of Anfinsen's dogma and subsequent theoretical considerations expounded above.

4.4   Membrane Proteins

In recent years, the three-dimensional structures of several membrane proteins have been determined: for example, bacteriorhodopsin in the purple membrane of *Halobacterium halobium* by Henderson *et al.*,[147] photosynthetic reaction center of *Rhodopseudominas viridis* by Deisenhofer *et al.*[148] and porin of *Rhodobacter capsulatus* by Weiss *et al.*,[149] etc. Membrane proteins are usually insoluble in water and their conformations are quite different from water-soluble globular proteins. The distinction between the globular and membrane proteins by means of their amino acid sequences was proposed by Yanagihara *et al.*[150] through the distribution of hydrophobic residues on the chain. They assigned the hydrophobic index to every amino acid residue and considered the sequence of hydrophobic indices. The average value and the periodicity of the hydrophobic indices are key parameters for their method. Their score of prediction amounts almost to 98%. The primary structure of bacteriorhodopsin which has 248 amino acid residues, was determined by Ovchinnikov *et al.*[151] and Khorana *et al.*[152] in 1979, and ten years later, its three-dimensional structure was elucidated at the atomic level by electron diffraction by Henderson *et al.*[153] In the membrane, it has 7 α-helices, among which three are almost perpendicular to the membrane and four others are at an incline about $20°$ to the normal of the membrane. The six loops connecting the seven helices and two coil parts at either end are immersed in the aqueous regions outside the membrane. The structures of the loops were not determined at that time (recently they were determined by Kimura et al.[154]). On the other hand, the photosynthetic reaction center is composed of L, M, H subunits, and cytochrome c. The three-dimensional structure together with the loops has been determined.[155] L and M subunits are similar and each of them has 5 helices penetrating the membrane. The 5 helices (say A, B, C, D, and E from the N-terminus) are almost perpendicular to the membrane surface. H subunit has 1 penetrating helix.

How this kind of structure can be formed is a challenging subject and indeed, Suwa et al.[156] tried to assemble the seven transmembrane helices in bacteriorhodopsin by searching for the mutual orientation of least energy, considering mainly the polar interaction. To do this, they had to calculate interaction energy for all the pairs of helices and then assembled them into the arrangement of least energy. In this treatment, however, the role of the loops connecting two neighboring helices was not taken into account. The tertiary structures of the loops restrict the movements of the neighboring helices and can certainly contribute to the arrangement of helices and help the theoretical computer-aided method of arranging helices in the membrane. Thus, an attempt to refold the tertiary structures of the loops of photosynthetic reaction center would be interesting. This is an application of the island model developed hitherto, but some modifications are required to be adapted to the restriction of the presence of membrane, such as the inhibition of free movement of the loop chain or some possibility to make contact with the membrane surface. The studies along this line are now in progress and will be presented elsewhere.

4.5   Structure Prediction Based on Protein Data

This section is a digression from the main stream of the present review, but a brief mention of this subject may be necessary, because the increasing accumulation of protein data in the Protein Data Bank (PDB) is developing a new field of protein research, which will meet the requirement of prompt determination of tertiary structures of many proteins whose primary structures become available from DNA sequence determination.

4.5.1 *Secondary structure*

Since it was recognized that the protein structure was determined solely by its amino acid sequence, many attempts were proposed to correlate the secondary structures in proteins with the amino acid sequence as early as in 1960s. Among them those by Lewis *et al.*[157,158] are worth mentioning. First the α-helical regions were considered, based on the helix-coil transition theory by Zimm and Bragg[28] that has two parameters s and σ which imply nucleation and helix formation respectively and correspond to v (LR theory) or w_0 (ours) and to v^2w (LR theory) or $w_0{}^3w_1^2w_2$ (ours) (see Appendix A). It is assumed that the nucleation parameter s is independent of amino acid types and temperature, while σ is strongly dependent on both. The s value for each amino acid was determined only approximately, but the resulting predictions of α-helical regions for several proteins were rather good.

In the β-bends Lewis *et al.*[158] assembled the data of amino acid sequences in the four residues forming the bend structure of various proteins and then, they estimated the probabilities of occurrence of an amino acid residue at the four positions of a bend. In this way, the probability of the formation of a bend for a sequence of four residues in a protein can be estimated with success. These attempts encouraged much subsequent research in two ways, *ab initio* method and comparative or homology method (see Sec. 1.2.5). See as an extensive collection of various researches on secondary structure prediction, Rost, Sander, and Schneider,[159,161] and the introduction of the paper by Rost and Sander.[113]

The statistical mechanical theory was extended to include the interaction of neighboring residues by Wako *et al.*[30] for two states (α-helix and coil, or β-strand and coil) and by Saitô[10] for three states (α-helix, β-strand and coil), as described in Sec. 1.2.

On the other hand, the comparative methods using protein data were also developed. For example, Chou and Fasman[162] determined the propensity indices of amino acids for helix, β-sheet, and β-turn to predict the secondary structures. The doublet and triplet propensities were also considered. Furthermore, evolutionary information of protein family were also taken into account. For the purpose of performing an appropriate processing of the information obtainable from protein data, multilayered neural networks were used. Thus, Rost and Sander[113] attained an accuracy better than 70%. We do not enter into this subject any further, but this trend led towards the application of the tertiary structure prediction.

4.5.2 *Tertiary structure*

The PDB contains two kinds of data, primary structures and tertiary structures. The number of latter data now almost reaches to 1.3×10^4 which is, nevertheless, much smaller than that of the former, but it is often sufficient to deduce the tertiary structures of proteins whose primary structures only are available. The main reason is that homologous proteins have similar tertiary structures, a fact revealed by the accumulation of protein data. This method is called homology recognition or comparative modeling, because it consists of comparing the primary structure of a protein of unknown structure with those of known structures and thus, it is not effective for a protein of low sequence homology with any protein of already known structure. Structural similarities are sometimes detected among the proteins of low sequence homology related distantly in molecular evolution. An example is the myoglobin of sperm whale and sea hare, which have the same biological function but low sequence homology as discussed in Sec. 2.2.3. Actin and heat shock protein HSP70[163] also have the same function of ATP binding and thus have similar structures retained almost unchanged during molecular evolution. Another interesting example is colistin which has a similar structure with, but is supposed to have a different origin from, myoglobin.[164] Furthermore, Chothia[165] maintained that the number of structures essentially different from each other is at most 1000. These facts suggest a new approach to predicting the tertiary structure of a protein. The method is to test the compatibility of a sequence of amino acid residues with the tertiary structures of known representative proteins. This is called the fold recognition method or the threading method and has attracted much attention, but it cannot predict a new structure the sequence may have, that is not similar to any protein already investigated.

4.6 Concluding Remarks

We have presented a detailed analysis of protein architecture. We believe that the main features of protein folding are clarified to the extent that it is close to being applicable to predicting the tertiary structure of a protein from its primary structure. The doorway has been opened to the second deciphering of genetic information. However, there are still several problems.

(1) *Assumptions or conjectures not fully verified.* One of them is that the secondary structures, especially in the nascent β-strands, are the standard

structures and not always the ones in those of the native structures, which are modified at the final stage of folding. This conjecture is partly verified in sea hare myoglobin and ferredoxin discussed in Sec. 2.2.3 and Sec. 2.6 respectively but full verification is necessary for the secondary structure prediction, as well as for detailed elucidation of the folding mechanism of proteins. Another assumption is that the structure of a protein in water is the same as the one in crystalline state. This was verified for lysozyme as mentioned in Sec. 1.1. This problem is related to the next discussion.

(2) *Effect of solvent or water surrounding protein.* A short discussion was given on the effect of solvent to the stability of an α-helix in Sec. 1.2.6, but other than this no mention was made on the effect of solvent except for the hydrophobic interaction. On this subject, Oobatake and Ooi,[166] in particular, considered the accessible surface area (ASA) that makes contact with the solvent molecules. Honig's group,[33,167] besides a simplified treatment of ASA, takes account of electrostatic interaction through Finite Difference Poisson-Boltzmann equation (FDPB/γ). Another statistical mechanical method called RISM (Reference Interaction Site Model) has also been developed.[168] These methods may yield slight differences of the structures between in crystal and in solution, but as far as the authors are aware, no attempts have been made yet to estimate the differences. The development of NMR technique will be required to promote this sort of research, as in the case of lysozyme (see Sec. 1.1).

(3) *Effect of temperature.* No explicit consideration was given in the main discussion of the protein architecture upto here. The temperature is always assumed to be room temperature or one suitable for living organisms. The parameters employed in the formulation are those of surrounding environments. The effect of temperature, however, is taken into account by the methods of simulated annealing or multicanonical algorithm (see Sec. 3.2.2, and Okamoto[169]). Consequently, for the purpose of considering proper accounts of temperature and solvent effects, one may have recourse to the method of simulated annealing in place of Bremermann's method for searching the conformation of least energy. This is also a task remaining for the future. It is noted, however, that the method of simulated annealing is not appropriate for application from the outset for obtaining the protein structure, as was already pointed out in Sec. 3.2.2.

Appendix A

Helix-Coil Transition in Homopolypeptide

A.1 Lifson-Roig Theory

In the case of homopolypeptide, the theory developed in Sec. 1.2.3 can be reformulated into an analytic explicit form. Now, we ignore the formation of β-strands. The function H introduced there depends only on k. On the other hand, in the Lifson-Roig (LR) model where only the interaction arising from hydrogen bondings is taken into account. p is put to 2 in Eq. (1.7) and the statistical weights assigned to the conformations represented by the sequences of coil(c) and helical parts(h) are assumed, as shown in Table A.1. In this table, the second column indicates the function H_k ($k = 1, 2, 3, \ldots$) corresponding to the conformations in the first column. We can identify $w_0 = v, w_1 = 1, w_0 w_2 = w$, and $w_3 = w_4 = \cdots = 1$. Thus, we have Fig. A.1 for the relation between helix fraction θ and $\ln w$. For large n, θ changes steeply from

Table A.1. Statistical weights.

Conformation	Our definition	Statistical weight	
		LR	Ours
chc	H_1	v	w_0
chhc	H_2	v^2	$w_0^2 w_1$
chhhc	H_3	$v^2 w$	$w_0^3 w_1^2 w_2$
..	..	..	..
chh $\cdots$ hc	H_k	$v^2 w^{k-2}$	$w_0^k w_1^{k-1} w_2^{k-2} \cdots w_{k-1}$

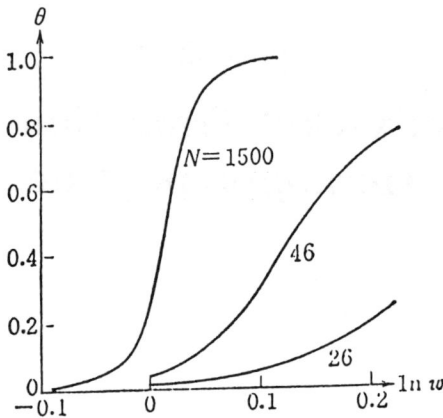

Fig. A.1.   Helix fraction θ versuss $\ln w$.

almost 0 to almost 1 at around $w = 1$. This also shows a sigmoidal diffuse transition owing to the cooperative nature of the interaction. For small n, the change in θ and also the change in the heat content are broad, but still the cooperativity lies behind the changes. These changes are sometimes found in molten globule states. The formalism discussed above can be applied to the formation of β-strands, by setting u_0 and $u = u_0 u_2$ for w_0 and $w = w_0 w_2$. In the case of a helix, the quantity w of nonzero value arises from a small region of dihedral angles due to the short-range hydrogen-bond interaction. However, in the case of a β-strand, the region of nonzero u is supposed broad. This fact gives rise to the instability of the β-strand compared with the α-helix.

A.2   Interaction of Side Chains

Now we can take the interactions between side chains into consideration. Look at Fig. 1.6. The side chain of the ith residue (ith side chain) can interact with the $(i + 1)$th through the $(i + 4)$th side chains. These interactions contribute to the statistical weights w_1, through w_4. Lifson and Roig set $w_1 = 1$ and considered only the hydrogen bonds between $C_i = O$ and H–N_{i+3}, which contribute to the statistical weight $w_2(i, i + 2)$, as mentioned in Sec. 1.2.3. In order to consider the statistical weights from w_1 to w_4, we set $p = 4$ in Eq. (1.7) and let w_2 include the statistical weight of the interaction

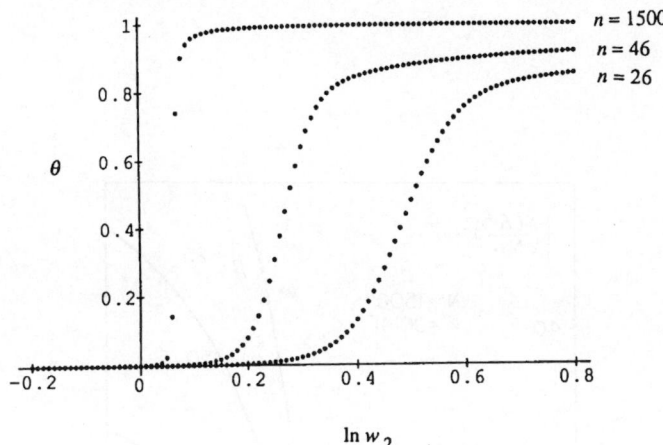

Fig. A.2. Helix fraction θ in a complicated model.

between the ith and the $(i + 2)$th side chains. Then, by tentatively setting $w_0 = 0.0141, w_1 = 3, w_3 = 5, w_4 = 4.5$, we finally obtain the relations between q and $\ln w_2$, which are more steep than those of Fig. A.1, as shown in Fig. A.2.

A.3 Conformation in the Helix-Coil Transition Region

The conformation in the coil state is of course a random coil, while it is rod-like in the helix state. The over-all conformation in the transition region is rather complicated, especially in electrolytic PGA as is seen in the value of the intrinsic viscosity in Fig. 1.4. Theoretical calculations of the root mean square end-to-end distance, which is related to the intrinsic viscosity, by Gō *et al.* [170] are shown in Fig. A.3.

The helix-coil transition is not the first order transition as in N-D transition in protein, as described in Sec. 1.3.

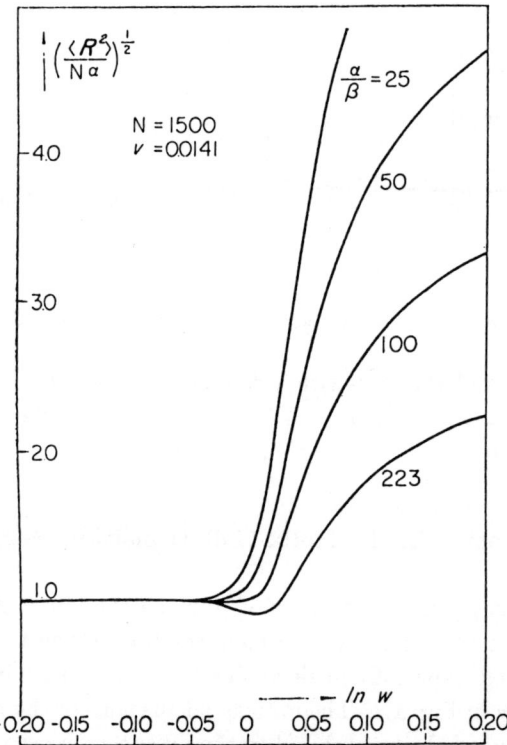

Fig. A.3.   Root mean square end-to-end distance versus $\ln w$ for various α/β values, where α and β are respectively the mean square end-to-end distances per residue for random coil state and rod-like helical state. N is the number of residues. α/β is large in polyelectrolytes (reproduced from Ref. 170 with permission).

Appendix B

Levinthal Paradox

B.1 Phase Space of a Protein

In Sec. 1.3.1 the first and second principles of protein folding were presented; the first one states that the conformation of a protein is governed by equilibrium thermodynamics and the second insists on the quickness of the formation of the tertiary structure. Levinthal cast a doubt on the applicability of conventional statistical thermodynamics by calculating the number of complexions in the conformation space of a protein, which amounts to 3^{2n} in a protein of $(n + 1)$ amino acid residues (Sec. 1.3.1). He pointed out that this number is formidably big to survey all the conformations in a short time, so that search for the state of minimum energy is carried out not over the whole conformation space, but over a restricted space, thus reconciling to the second principle. This restriction on the conformation space is really done in protein tertiary structures, as shown in the whole pages of the text through the formation of relevant hydrophobic bindings. In the treatment of the formation of the secondary structures discussed in Sec. 1.2.3, only the interactions of those necessary for forming α-helices or β-strands are considered and thus one does not have to consider the space where the long distance interaction might take place. Consequently the conformation space is largely restricted. But one may raise a question about the case of random coil homopolymers of n degrees of polymerization, where the number of complexions in the conformation space is almost of the same order as the proteins of n residues, but in constructing theories for solution properties or for statistical properties of a single molecule, no attempt is made to restrict the phase space. To answer this question, we shall discuss the case of an ideal gas first.

B.2   Ideal Gas

Now, consider a gas of volume V and energy E, composed of N monatomic molecules. Since the energy is written as

$$\sum_{i=1}^{N} \frac{1}{2m}(p_x{}^2 + p_y{}^2 + p_z{}^2) = E,$$ (B.1)

the volume of the momentum space is equal to the surface area of the $n = 3N$ dimensional sphere of radius $r = \sqrt{2mE}$,

$$\frac{n\pi^{n/2}}{(\frac{n}{2})!} r^{n+1} = \frac{1}{(\frac{3}{2}N)!} \frac{3N}{\sqrt{2mE}} (2\pi mE)^{\frac{3}{2}N}.$$ (B.2)

On the other hand, since the volume of the configuration space is V^N, the number of complexions are

$$W = \frac{1}{N! h^{3N}} \frac{V^N}{(\frac{3}{2}N)!} \frac{3N}{\sqrt{2mE}} (2\pi mE)^{\frac{3}{2}N}.$$ (B.3)

where $N!$ is introduced in the denominator due to the identical particles and h^{3N} is the size of the cell of one state. By making use of Stirling's approximation, Eq. (B.3) turns out to be

$$W = \frac{1}{\sqrt{2mE}} \sqrt{\frac{3}{2} \frac{1}{\pi}} \left[\left(\frac{V}{N} e\right) \left(\frac{4\pi mE}{3h^2 N} e\right)^{3/2} \right]^N$$ (B.4)

where V/N and E/N are respectively the volume and the energy per molecule and they can take arbitrary values. Consequently, the factor in the parenthesis $[\cdots]$ can take a value bigger than 1, and Eq. (B.4) becomes an extraordinarily large number for $N \sim 10^{24}$ that can prohibit the itineration over the states in a reasonable time. Nevertheless, the conventional statistical mechanics holds without restricting the phase space. Why?

This is because the important factor in Eq. (B.4) is of type $[\cdots]^N$. This gives rise to the extensive property of the extensive thermodynamical quantities, such as volume, energy, and entropy, etc. The same holds for the partition function Z of the canonical ensemble

$$Z = \frac{1}{N!}(Z')^N \cong Z_1^N,$$ (B.5)

$$Z' = (2\pi mkT)^{3/2}V,$$ (B.6)

$$Z_1 = e\frac{V}{N}(2\pi mkT)^{3/2}$$ (B.7)

where Z' is the partition function of one particle in the volume V. Z_1 is the partition function of a particle in the volume V/N allotted to one particle and the entropy arising from the factor e is sometimes called communal entropy. Thus, in the calculation of Z, one does not have to consider Z itself of the large V and N, but only to consider Z_1 and a much smaller volume V/N, which allows to itinerate over the small phase space in a reasonable amount of time. Here is the legitimacy of the statistical mechanical treatment of gases. Generally speaking, the relation of (B.5) implying the extensive property of the system guarantees this treatment. In a system composed of interacting identical particles, we take a particle and properly consider the partition function involving this particle and the interaction with others. Since the system is large enough, this partition function is the same even if we take another particle. Thus, we have an extensive property similar to the relation (B.5). These considerations yield the conventional treatment of statistical mechanics mentioned at the end of the previous paragraph. Almost the same argument holds for homopolymers, where one has only to consider the end effect when necessary.

Appendix C

Method of Bremermann for Searching the Conformation of Minimum Energy

C.1 Outline of the Method

Following Yčas *et al.* [86], we search the conformation of minimum energy by the optimization method developed by Bremermann.[85] The procedure is shown in Fig. C.1.

(1) A set of dihedral angles ϕ and ψ of the residues in the coil part is chosen and a multi-dimensional space spanned by these angles is considered. The dihedral angles ϕ and ψ can be changed in the range of $-180°$ to $180°$.

(2) The energy is evaluated for an initial point X_0, which represents an extended conformation with all dihedral angles of coil parts set to $180°$ at the initial stage of packing or an intermediate conformation during the course of packing.

(3) A unit vector r of random direction is chosen in the space by generating random numbers. Consider the line ℓ: $X_0 + \lambda r$ in Fig. C.2 and regard the line ℓ as 'an axis of coordinates', and ℓ as the variable of the coordinate ($\ell = 0$ at the origin). Thus, the energy is expressed in terms of one variable λ.

(4) Four different values of $\lambda_i (i = 1, \ldots, 4)$ are chosen along the line ℓ and the energies are estimated at these points.

(5) The objective function, $F(\lambda)$, is assumed of the form $a\lambda^4 + b\lambda^3 + c\lambda^2 + d\lambda + e$, where the coefficients are determined by the energies estimated at the 5 points X_0 and $X_i = X_0 + \lambda_i r (i = 1, \ldots, 4)$.

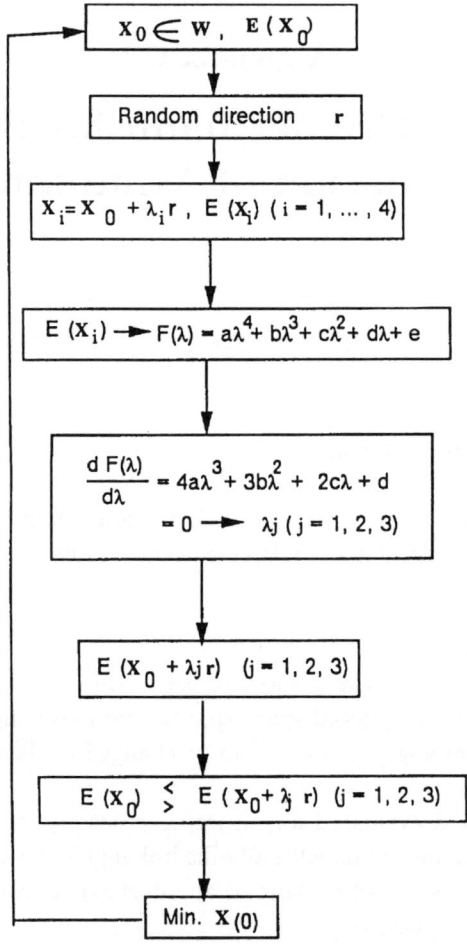

Fig. C.1.   Procedure of energy minimization.

(6) The minima of $F(\lambda)$ are searched along the line ℓ (Fig. C.3). They are estimated from the derivative of $F(\lambda)$, which is a cubic polynomial and thus has one or three real roots.

(7) The least minimum of $F(\lambda)$ is compared with $F(0)$. It is by no means bigger than $F(0)$. Then we move the conformation from X_0 to the one, say X_1, having the least value of $F(\lambda)$.

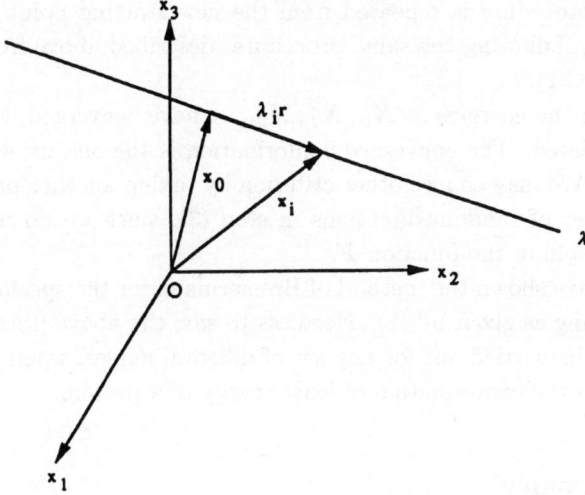

Fig. C.2. Multi-dimensional space spanned by the dihedral angles of coil parts. For simplicity, the three-dimensional space is shown. In treating n dihedral angles as variables, we consider the n-dimensional space spanned by these dihedral angles.

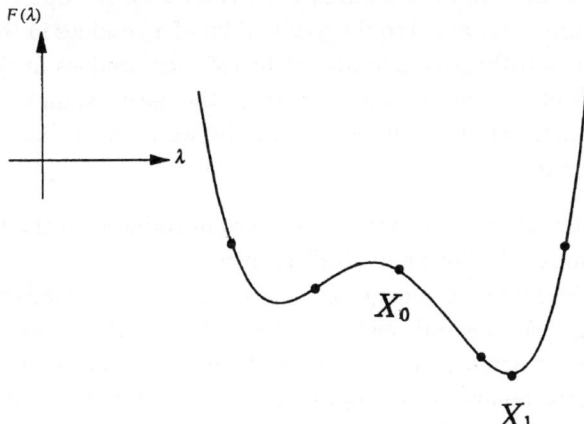

Fig. C.3. Search for the minimum energy by minimizing the objective function $F(\lambda)$. The energy is expressed in terms of one parameter λ, which corresponds to the coordinate on the "axis" with a random direction in the space. The first search of minimum energy is performed along a line with random direction. The next search is carried out along a line with another random direction. This procedure is iterated until the predetermined number of random directions is chosen.

(8) The procedure is repeated from the new starting point X_1 in stead of X_0, following the same procedures described above from (2) to (7) (Fig. C.1).

(9) When the energies at $X_0, X_1, X_2, \ldots$ have converged, the process is completed. The converged conformation is the one we wanted to obtain. We may choose other criterion by taking another predetermined number of random directions in step (3), since we do not know the threshold of the function F.

(10) We have shown the method of Bremermann for the special set of dihedral angles given by (1). Needless to say, the above iteration process must be carried out for any set of dihedral angles, when necessary, to obtain the conformation of least energy of a protein.

C.2   An Example

We show the process of energy minimization by choosing cytochrome b_{562} as an example. The process is as follows:

(1) The extended conformation is built (Fig. 2.7), in which the dihedral angles are constrained to the native values for residues in the secondary structures (α-helices) and are set to $180°$ for residues in the coil parts (in case of proline, ϕ is constrained to the native value).

(2) The conformation of the chain part between two helices A and B is considered.

 (2a) The hydrophobic interactions are introduced to the hydrophobic pair of the 20th and 23rd residues.

 (2b) The minimum energy of this chain part is searched by changing the dihedral angles of the 21st and 22nd residues in the 4-dimensional pace spanned by these angles. The starting point X_0 corresponds to the initial extended conformation. At the earlier stage of packing, small changes in the dihedral angles tends to bring about large changes of the whole conformation.

 (2c) The interactions are introduced to the hydrophobic pairs that are remote along the chain.

 (2d) The energy is minimized by choosing the conformation obtained at step (2b) as the starting point X_0 and changing the dihedral angles of the 20th, 21st, and 22nd residues in the 6-dimensional space.

(2e) When the expected hydrophobic pairs are bound, we proceed to the next step.

(3) The conformation of the chain part between two helices C and D is considered.

(3a) The hydrophobic interactions are introduced to the hydrophobic pairs at intervals of 5 residues along the chain.

(3b) The energy is minimized by changing the dihedral angles of the 81st, 82nd, and 83rd residues in the 6-dimensional space. The starting point X_0 corresponds to the conformation obtained at step 2.

(3c) When the expected hydrophobic pairs are bound in step 3, we proceed to step 4.

(4) The conformation of the remaining parts is considered.

(4a) In the above procedures, the interactions are considered from nearer pairs to the remote ones along the chain (i.e. from short distance to longer ones) and the energy is minimized at each stage.

(4b) The energy minimization is completed, when the target hydrophobic pairs, which play an important role for packing, are bound. Even when the interaction range is extended, the whole conformation is kept almost unchanged, despite the small change in the dihedral angle. We expect that step-by-step refolding along the above definite folding pathway can avoid the multi-minima problem sometimes considered to be one of the difficulties in predicting protein conformation.

Appendix D

Formation of β-Sheets

D.1   Antiparallel β-Structure

The formation of secondary structures from a nascent random-coil polypeptide is very quick, since they are constructed from the interactions of nearer residues. The secondary structures thus established contribute the main features of the molten globule state, as explained in Secs. 1.3.3 and 1.3.4. The antiparallel β-structures are supposed to be formed between the nearest β-strands. The interaction responsible for the formation of this structure will be found by looking at the conformations of known structures with the purpose in mind to obtain some general rules. To begin with, we shall try to search for them in lysozyme and phospholipase.[14] Lysozyme has five β-strands (see Sec. 2.3.1).

Among them, a β_1 strand, that is rather separated from other β-strands, makes contact with an α-helix. We have only to consider β_2 through β_5. They are shown in Fig. D.1(a). On the other hand, phospholipase has two β-strands lying side by side, as shown in Fig. D.1(b). It may be easily assumed that hydrophobic interaction plays a role, as in the packing of secondary structures (α-helices and β-sheets) into tertiary structures in a way described in Chapters 2 and 3. In the native structure of lysozyme, antiparallel β-strands are formed between β_3 and β_4 and between β_4 and β_5 as shown in Fig. D.1(a). One immediately sees that Phe 38 and Ile 58 are the only strong hydrophobic residues (see Sec. 1.3.2) in these β-strands. Hence, weak hydrophobic residues Tyr, Cys, and Ala has to be taken into consideration. Now, we can understand the formation of the antiparallel β-structure between β_4 and β_5 in lysozyme due

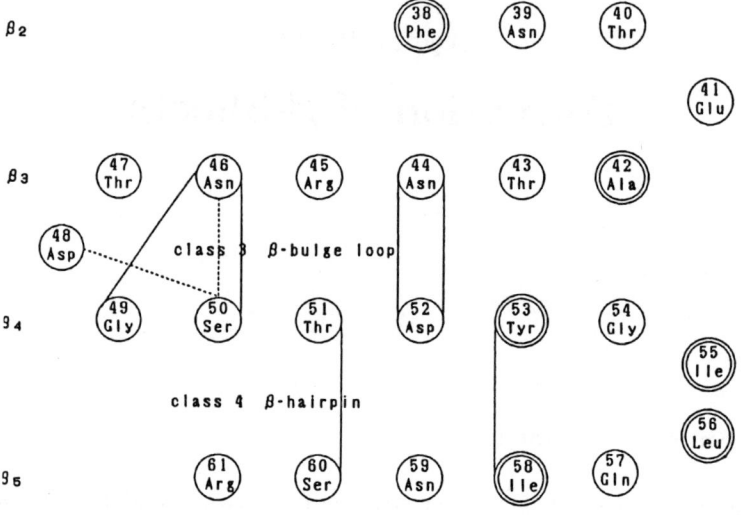

(a) LYSOZYME (HEN EGG WHITE) 6LYZ

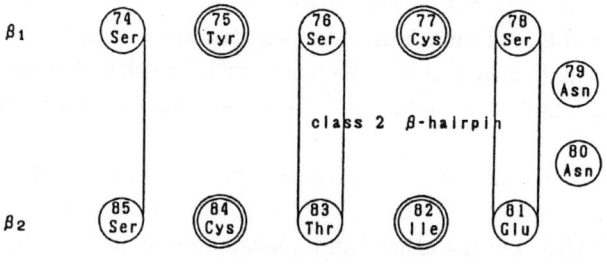

(b) PHOSPHOLIPASE 1BP2

Fig. D.1. β-structures in lysozyme and phospholipase. ———: hydrogen bond between main chain atoms, ······: hydrogen bond between side chains, ⊙: hydrophobic residue.

to the interaction between Tyr53 and Ile58, as illustrated in Fig. D.1(a). The same holds for phospholipase. The pairs of Cys77 and Ile82 and of Tyr75 and Cys84 can be bound by the hydrophobic interaction, as shown in Fig. D.1(b). In these cases the hydrophobic residues are located on β-strands on the part close to β-turns, and when they come close, zipping of hydrogen bonding can take place. It is also to be noted that the β-turns are class 2 β-hairpins (i.e. 2 residues at the β-turn) in phospholipase and class 4 β-hairpins (4 residues at the β-turn) in lysozyme in the terminology of Milner-White and Poet.[171] This difference may be attributed to the presence or absence of a hydrophobic interaction at the turn. In lysozyme, two hydrophobic residues (Ile55 and Leu56) are situated side by side at the turn. The antiparallel β-structure between β_3 and β_4 in lysozyme, however, cannot be explained by the hydrophobic interaction. The β-turn between β_3 and β_4 is a class 3 β-hairpin loop with a hydrogen bond between CO of Asn46 and N of Gly49. There is also a hydrogen bond between N $=$ O of the side chain of Asn48 and O of the side chain of Ser50 beyond Gly49. The two hydrophobic residues, Ala42 and Tyr53, are at the far opposite ends of the two β-strands and are not easily bound to yield an antiparallel β-structure of class 2 or class 4 β-hairpin, which might be expected to occur just as between β_4 and β_5. The class 3 β-hairpin between β_3 and β_4 is formed by the two hydrogen bonds mentioned above. No antiparallel β-structure can be expected between β_2 and β_3 because there is no driving force for this structure. This is in fact the case, as one can see in Fig. 2.20 in the text. The mechanism of the formation of an antiparallel β-structure, however, has not yet been completely elucidated. There are some examples which do not fall into the categories mentioned above. In particular, *cis-trans* transformation in proline observed, for example, in ribonuclease A is supposed to be related to the formation of antiparallel β-structures. In fact, *cis* Pro93 and 114 of ribonuclease A are situated respectively at the turn of β_3 and β_4, and at the turn of β_4 and β_5.

References

1. See, as text books of DNA biology and technology, for example, M. Singer and P. Berg, *Genes and Genomes — A changing perspective* (University Science Book, 1991). B. Jordan, *Traveling around the human genome* (IN-SERM John Library Eurotext, 1993) (Japanese translation by S. Mitaku, Kodansha, 1995).
2. F. Sanger and A. R. Coulson, J. Mol. Biol. **94**, 441 (1975).
3. A. M. Maxam and W. Gilbert, Proc. Natl. Acad. Sci. **74**, 560 (1977).
4. L. J. Smith, M. J. Sutcliffe, C. Redfield and C. M. Dobson, J. Mol. Biol. **229**, 830 (1993).
5. J. Kobayashi, T. Asahi, M. Sakurai, I. Kagomiya, H. Asai and H. Asami, Acta Cryst. **A54**, 581 (1998).
6. N. Saitô, H. Wako, T. Akutsu and Y. Oyama, in Ions and Molecules in Solution, Proc. VI, ISSSSI, Minoo, Osaka, July 4–10 (Elsevier Publ. Co., 1982).
7. N. Saitô, Mem. School Sci. Eng. Waseda University, **46**, 295 (1992).
8. N. Saitô, in *Biomolecules, Electronic Aspects* (ed. C. Nagata, J. Tanaka and H. Suzuki, Japan Sci. Soc. Press and Elsevier, 1985), 53.
9. N. Saitô, in *Computer Analysis for Life Science* (eds. C. Kawabata and A. R. Bishop, Ohmsha, 1986), 22.
10. N. Saitô, Cell Biophys. **11**, 321 (1987).
11. N. Saitô,T. Shigaki, Y. Kobayashi and M. Yamamoto, Proteins Struct. Funct. Genet. **2**, 199 (1988).
12. N. Saitô, K. Yura and Y. Fukuda, in *Protein Structural Analysis, Folding and Design* (ed. M. Hatano, Japan Sci. Soc. Press and Elsevier, 1990), 19.

13. N. Saitô, Adv. Biophys. (ed. M. Kotani, Japan Sci. Soc. Press and Elsevier, 1989) **25**, 95.

14. T. Yoshimura, H. Noguchi, T. Inoue and N. Saitô, Biophys. Chem. **40**, 277 (1991).

15. K. Watanabe, A. Nakamura, Y. Fukuda and N. Saitô, Biophys. Chem. **40**, 293 (1991).

16. Y. Kobayashi, H. Sasabe, T. Akutsu and N. Saitô, Biophys. Chem. **44**, 113 (1982); Errata, *ibid.* **45**, 195 (1992).

17. N. Saitô, Y. Kobayashi, M. Ota and S. Mitaku, Rep. Prog. Poly. Phys. Jpn. **35**, 1 (1992).

18. Y. Kobayashi, H. Sasabe and N. Saitô, J. Protein Chem. **12**, 121 (1993).

19. M. Ota and N. Saitô, J. Protein Chem. **11**, 623 (1992).

20. Y. Kobayashi and N. Saitô, J. Protein Chem. **16**, 83 (1997).

21. L. Pauling, R. C. Corey and H. R. Branson, Proc. Natl. Acad. Sci. USA **37**, 205 (1951); L. Pauling and R. C. Corey, Proc. Natl. Acad. Sci. USA **37**, 235 (1951).

22. W. Kabsch and C. Sander, Biopolymers **22**, 2577 (1983).

23. P. Doty, Rev. Mod. Phys. **31**, 107 (1959); P. Doty, A. Wada, T. Yang and E. R. Blout, J. Polymer Sci. **23**, 851 (1957).

24. M. Toda, R. Kubo and N. Saitô, *Statistical Physics* (second ed.) Vol. 1, Chapter 4 (Springer, 1995).

25. N. Saitô and M. Oonuki, J. Phys. Soc. Jpn. **27**, 69 (1970).

26. D. Poland and H. A. Scheraga, *Theory of Helix-Coil Transition in Biopolymers* (Academic Press, New York and London, 190).

27. T. L. Hill, J. Chem. Phys. **30**, 358 (1959).

28. B. H. Zimm and J. K. Bragg, J. Chem. Phys. **31**, 526 (1959).

29. S. Lifson and A. Roig, J. Chem. Phys. **34**, 1963 (1961).

30. H. Wako, N. Saitô and H. A. Scheraga, J. Protein Chem. **2**, 221 (1983).

31. K. Kosuge, M. Fujiwara, Y. Isogai and N. Saitô, Polymer J. **4**, 100 (1973).

32. T. Ooi, R. A. Scott, G. Vanderkooi and H. A. Scheraga, J. Chem. Phys. **46**, 4410 (1967).

33. A.-S. Yang and B. Honig, J. Mol. Biol. **252**, 351, 366 (1995); A.-S. Yang, B. Hitz and B. Honig, J. Mol. Biol. **259**, 873 (1996).

34. E. Harber and C. B. Anfinsen, J. Biol. Chem. **237**, 1839 (1962); C. B. Anfinsen, Science **181**, 223 (1963).

35. C. B. Anfinsen and H. A. Scheraga, Adv. Protein Chem. **29**, 205 (1975).

36. T. Takagi and T. Isemura, J. Biochem. **52**, 314 (1962).

37. T. Isemura, in Biopolymers (ed. Biophysical Society of Japan, Yoshioka Shoten, Kyoto, 1965) **1**, 41.
38. T. Isemura, T. Takagi, Y. Maeda and K. Imai, Biochem. Biophys. Research Communication **5**, 371 (1961).
39. M. Ikeguchi, K. Kuwajima and S. Sugai, J. Biochem. **99**, 1191 (1986).
40. C. Tanford, Adv. Protein Chem. **23**, 121 (1968).
41. C. Levinthal, J. Chim. Phys. Physico-Chim. Biol. **65**, 44 (1968).
42. R. Jaenicke, Biophy. Struct. Mech. **8**, 231 (1982).
43. C. Levinthal, in *Mossbauer Spectroscopy in Biological Systems*, Proceedings of Meeting held at Allerton House, Monticello IL (eds. P. Debrunner, J. C. M. Tsibris and E. Münck, University of Illinois Press, 1969), 22.
44. W. Kauzman, Adv. Protein Chem. **14**, 1 (1959).
45. G. Némethy and H. A. Scheraga, J. Phys. Chem. **66**, 1773 (1962).
46. Y. Nozaki and C. Tanford, J. Biol. Chem. **246**, 2211 (1971).
47. D. D. Jones, J. Theor. Biol. **50**, 167 (1975).
48. J. Israelachvili and R. Pashley, Nature **300**, 341 (1982).
49. L. C. Wu and P. S. Kim, Proc. Natl. Acad. Sci. USA **94**, 14314 (1997).
50. R. Srinivasan and G. D. Rose, Proteins Struct. Funct. Genet. **22**, 81 (1995).
51. S. Sun, P. D. Thomas and K. A. Dill, Protein Eng. **8**, 769 (1995).
52. O. B. Ptitsyn, J. Protein Chem. **6**, 273 (1967).
53. K. Kuwajima, Proteins Struct. Funct. Genet. **6**, 87 (1989).
54. P. S. Kim and R. I. Baldwin, Annu. Rev. Biochem. **59**, 631 (1990).
55. K. A. Dill and D. Shortle, Annu. Rev. Biochem. **60**, 795 (1991).
56. M. Arai and K. Kuwajima, Adv. Protein Chem. **53**, 209 (2000).
57. S. Sugai and M. Ikeguchi, Adv. Biophys. (ed. S. Ebashi, Japan Sci. Soc. Press and Elsevier, 1994) **30**, 37.
58. A. F. Chaffotte, Y. Guillou and M. E. Goldgerg, Biochem. **31**, 6876 (1992).
59. K. Kuwajima, Y. Hiraoka, M. Ikeguchi and S. Sugai, Biochem. **24**, 874 (1985).
60. M. Ikeguchi, K. Kuwajima and S. Sugai, J. Biochem. **99**, 1191 (1986).
61. M. Ikeguchi, K. Kuwajima, M. Mitani and S. Sugai, Biochem. **25**, 6965 (1986).
62. M. J. Kronman, L. Cerankowski and L. G. Holmes, Biochem. **4**, 518 (1965).
63. K. Kuwajima, M. Mitani and S. Sugai, J. Mol. Biol. **206**, 547 (1989).

64. M. Ohgushi and A. Wada, FEBS Lett. **164**, 21 (1983).
65. M. Kataoka, K. Kuwajima, F.Tokunaga and Y. Goto, Protein Sci. **6**, 422 (1997).
66. J. Baum, C. M. Dobson, P. A. Evans and C. Hanley, Biochem. **28**, 7 (1989).
67. S. Mitaku, S. Ishido, Y. Hirano, H. Itoh, R. Kataoka and N. Saitô, Biophys. Chem. **49**, 217 (1991).
68. P. L. Privalov, Adv. Protein Chem. **33**, 167 (1979).
69. M. Ikeguchi, K. Kuwajima, M. Mitani and S. Sugai, Biochem. **25**, 6965 (1986).
70. K. Fujiwara, M. Arai, A. Shimizu, M. Ikeguchi, K. Kuwajima and S. Sugai, Biochem. **38**, 4455 (1999).
71. W. Colón, G. A. Elöve, L. P. Wakem, F. Sherman and H. Roder, Biochem. **35**, 5538 (1996).
72. W. Colón and H. Roder, Nature Struct. Biol. **3**, 1019 (1996).
73. S. F. Jackson and A. R. Fersht, Biochem. **30**, 10428 (1991).
74. M. Nozaka, K. Kuwajima, K. Nitta and S. Sugai, Biochem. **17**, 3753 (1978).
75. M. Ikeguchi, K. Kuwajima and S. Sugai, J. Biochem. (Tokyo) **99**, 1191 (1986).
76. M. Ikeguchi, M. Fujino, M. Kato, K. Kuwajima and S. Sugai, Protein Sci. **7**, 1564 (1998).
77. K. Yutani, K. Ogasawara and K. Kuwajima, J. Mol. Biol. **228**, 347 (1992).
78. D. C. Phillips, Biochem. Soc. Symp. **31**, 11 (1970).
79. K. Nishikawa, T. Ooi, Y. Isogai and N. Saitô, J. Phys. Soc. Jpn. **32**, 1331 (1972).
80. M. Gō, Nature **291**, 90 (1982).
81. H. Wako and N. Saitô, J. Phys. Soc. Jpn. **44**, 1931 (1978).
82. H. Wako and N. Saitô, J. Phys. Soc. Jpn. **44**, 1939 (1978).
83. A. Ikegami, Adv. Chem. Phys. **46**, 363 (1981).
84. F. A. Momany, R. F. McGuire, A. W. Burgers and H. A. Scheraga, J. Phys. Chem. **79**, 2361 (1975).
85. H. A. Bremermann, Math. Biophys. **9**, 1 (1970).
86. M. Yčas, M. S. Goel and J. W. Jacobson, J. Theor. Biol. **72**, 443 (1978).
87. B. R. Brooks, R. E. Bruccoleri, B. D. Olafson, D. J. States, S. Swaminathan and M. Karplus, J. Comp. Chem. **4**, 167 (1983).

88. W. H. Anderson and D. B. Wetlaufer, J. Biol. Chem. **351**, 3147 (1976).

89. H. Wakana, H. Yokomizo, H. Wako, Y. Isogai, K. Kosuge and N. Saitô, Int. J. Peptide Protein Res. **23**, 675 (1984).

90. H. Wakana, H. Wako and N. Saitô, Int. J. Peptide Protein Res. **23**, 315 (1984).

91. A. S. Acharya and H. Taniuchi, J. Biol. Chem. **251**, 6934 (1976).

92. A. S. Acharya and H. Taniuchi, J. Biol. Chem. **256**, 1905 (1980).

93. A. S. Acharya and H. Taniuchi, Mol. Cell. Biochem. **44**, 128 (1982).

94. T. E. Creighton, J. Mol. Biol. **95**, 167 (1975).

95. T. E. Creighton, J. Mol. Biol. **113**, 275 (1977).

96. T. E. Creighton, Prog. Biophys. Mol. Biol. **33**, 231 (1978).

97. T. E. Creighton and D. P. Goldenberg, J. Mol. Biol. **179**, 497 (1984).

98. D. J. States, T. E. Creighton, C. M. Dobson and M. Karplus, J. Mol. Biol. **195**, 731 (1987).

99. T. E. Creighton, Biochem. J. **270**, 1 (1990).

100. T. E. Creighton, *Proteins, Structure and Molecular Principles* (W. H. Freeman, New York, N.Y., 1983).

101. C. B. Marks, H. Naderi, P. A. Kosen, I. D. Kuntz and S. Anderson, Science **235**, 1370 (1987).

102. L. F. Kress and M. S. Laskowski, J. Biol. Chem. **242**, 4925 (1967).

103. R. Huber, D. Kukura, A. Rühlmann. O. Epp and H. Formanek, Naturwissenschaften **57**, 389 (1970).

104. N. J. Darby, C. P. M. Van Mierlo and T. E. Creighton, FEBS Lett. **297**, 61 (1991).

105. J. S. Weissman and P. S. Kim, Science **253**, 1386 (1991).

106. J. S. Weissman and P. S. Kim, Proc. Natl. Acad. Sci. USA **89**, 9900 (1992).

107. T. E. Creighton, Science **256**, 111 (1992).

108. D. P. Goldenberg, TIBS **17**, 257 (1992).

109. N. J. Darby, C. P. M. van Mierlo, G. H. E. Scott, D. Neuhaus and T. E. Creighton, J. Mol. Biol. **224**, 905 (1992).

110. N. J. Darby, P. E. Morin, G. Tarbo and T. E. Creighton, J. Mol. Biol. **249**, 463 (1995).

111. E. T. Adman, L. C. Sieker and L. H. Jensen, J. Biol. Chem. **298**, 3987 (1973).

112. K. Nishikawa and T. Noguchi, Meth. Enzymol. **202**, 31 (1991).

113. B. Rost and C. Sander, J. Mol. Biol. **232**, 584 (1993).

114. K. Nishikawa, H. Iwama and N. Saitô, *Computer Aided Innovation of New Materials* II (eds., M. Doyama, J. Kihara, M. Tanaka and R. Yamamoto, Elsevier Science Publishers B.V., 1993), 1271.

115. J. A. Barden and B. E. Kemp, Eur. J. Biochem. **184**, 379 (1989).

116. M. Vásquez, G. Némethy and H. A. Scheraga, Chem. Rev. **94**, 2183 (1994).

117. S. Sun, P. D. Thomas and K. A. Dill, Protein Eng. **3**, 85 (1995).

118. N. Metropolis, A. W. Rosenbluth, M. N. Rosenbluth, A. H. Teller and E. Teller, J. Chem. Phys. **21**, 1087 (1953).

119. S. Kirpatrick, C. D. Gelatt, Jr. and M. P. Vecchi, Science **20**, 671 (1983).

120. B. A. Berg, T. Celik and U. H. E. Hansmann, Phys. Rev. **B50**, 16444 (1994).

121. S. R. Wilson, W. Cui, J. W. Moskowitz and K. E. Schmidt, Tetrahedron Lett. **29**, 4373 (1988).

122. H. Kawai, T. Kikuchi and Y. Okamoto, Protein Eng. **3**, 85 (1989).

123. Y. Okamoto, T. Kikuchi, T. Nakazawa and H. Kawai, Int. J. Peptide Protein Res. **42**, 300 (1993).

124. S. R. Wilson and W. Cui, *The Protein Folding Problem and Tertiary Structure Prediction* **43** (eds. K. M. Merz, Jr. and S. M. Le Grand, Birkhauser, 1994).

125. B. A. Berg and T. Neuhaus, Phys. Lett. **B267**, 249 (1991); Phys. Rev. Lett. **68**, 9 (1992).

126. B. A. Berg, Int. J. Mod. Phys. **C**, 1083 (1992).

127. U. H. E. Hansmann and Y. Okamoto, J. Comp. Chem. **14**, 1333 (1993).

128. U. H. E. Hansmann and Y. Okamoto, Physica **A212**, 425 (1994); J. Phys. Soc. Jpn. **63**, 3945 (1994).

129. S. M. Berget, C. More and P. A. Sharp, Proc. Natl. Acad. Sci. USA **74**, 3171 (1977).

130. W. Gilbert, Nature **271**, 501 (1978).

131. E. F. Jensen, K. Paludan, J. J. Hyldig-Nielsen, PJørgensen and K. A. Marcker, Nature **291**, 677 (1981).

132. M. Gō, Adv. Biophys. (ed. M. Kotani, Japan Sci. Soc. Press and Elsevier, 1985) **19**, 91.

133. M. Gō and M. Nosaka, Cold Spring Harbor Symp. Quant. Biol. **52**, 915 (1987).

134. M.-Q. Xu, S. D. Kathe, H. Goodrich-Blair, S. A. Nierzwicki-Bauer and D. A. Shub, Science **250**, 1566 (1990).

135. M. G. Kushel, R. Strickland and J. D. Palmer, Science **250**, 1570 (1990).
136. J. E. Darnell and W. F. Doolittle, Proc. Natl. Acad. Sci. USA **83**, 1271 (1986).
137. T. Ikura, N. Gō, D. Kohda, F. Inagaki, H. Yanagawa, M. Kawabata, S. Kawabata, S. Iwanaga, T. Noguchi and M. Gō, Proteins Struct. Funct. Genet. **16**, 341 (1993).
138. T. Noguchi, H. Sakakibara and M. Gō, Proteins Struct. Funct. Genet. **16**, 357 (1993).
139. R. A. Laskey, B. M. Honda, A. D. Mills and J. T. Finch, Nature **275**, 416 (1978).
140. K. Braig, Z. Otwinowski, R. Hegde. D. C. Boisvert, A. Joachimiak, L. Horwich and P. B. Sigler, Nature **371**, 578 (1994).
141. J. F. Hunt, A. J. Weaver, S. J. Landry, L. Gierasch and J. Deisenhofer, Nature **379**, 37(1996).
142. S. C. Mande, V. Mehra, B. R. Bloom and W. G. J. Hol, Science **271**, 203 (1996).
143. M. Mayhew, A. C. R. da Silva J. Martin, H. Erdjument-Bromage, P. Tempst and F. U. Hartl, Nature **379**, 420 (1996).
144. J. S. Weissman, C. M. Hohl, O. Kovalensko, Y. Kashi, S. Chen, K. Braig, H. R. Staibil, W. A. Fenton, J. M. Beeche and A. L. Horwich, Cell **83**, 577 (1995).
145. J. S. Weissman, H. S. Rye, W. A. Fenton, J. M. Beeche and A. L. Horwich, Cell **84**, 481 (1996).
146. W. A. Fenton, J. S. Weissman and A. L. Horwich, Chem. Biol. **3**, 157 (1996).
147. R. Henderson, J. M. Baldwin, T. A. Cheka, F. Zemlin, E. Beckman and K. H. Downing, J. Mol. Biol. **213**, 899 (1990).
148. J. Deisenhofer, O. Epp, K. Miki, R. Huber and H. Michel, Nature **318**, 618 (1985).
149. M. S. Weiss, T. Wacker, J. Weckesser, W. Welte and G. E. Shults, FEBS Lett. **280**, 379 (1991).
150. N. Yanagihara, M. Suwa and S. Mitaku, Biophys. Chem. **34**, 69 (1989).
151. Yu A. A. Ovchinnikov, M. Yu Feigina, A. Kiselev and N. A. Lobanov, FEBS Lett. **100**, 219 (1979).
152. H. G. Kohrana, G. E. Gerber, W. C. Herlihy, C. P. Grey, R. J. Anderegg, K. Nihel and K. Biemann, Proc. Natl. Acad. Sci. USA, **76**, 5046 (1979).

153. R. Henderson, J. M. Baldwin, T. A. Cheka, F. Zemlin, E. Beckman and K. H. Downing, J. Mol. Biol. **213**, 899 (1990).

154. Y. Kimura, D. G. Vassylyev, A. Miyazawa, A. Kidera, M. Matsushima, K. Mitsuoka, K. Murata, T. Hirai and Y. Fujiyoshi, Nature **89**, 206 (1997).

155. J. Deisenhofer, O. Epp, K. Miki, R. Huber and H. Michel, Nature **318**, 618 (1995).

156. M. Suwa, T. Hirokawa and S. Mitaku, Proteins Struct. Func. Genet. **22**, 363 (1995).

157. P. N. Lewis, N. Gō, M. Gō, D. Kotelchuck and H. A. Scheraga, Proc. Natl. Acad. Sci. USA **65**, 810 (1970).

158. P. N. Lewis, F. A. Momany and H. A. Scheraga, Proc. Natl. Acad. Sci. USA **68**, 2293 (1971).

159. B. Rost, C. Sander and R. Schneider, Trends Biochem. Sci. **18**, 120 (1993).

160. B. Rost and C. Sander, J. Mol. Biol. **232**, 584 (1993).

161. B. Rost, C. Sander and R. Schneider, J. Mol. Biol. **235**, 13 (1994).

162. P. Y. Chou and U. D. Fasman, Biochem. **13**, 211 (1974).

163. K. M. Flaherty, D. B. McKay, W. Kabsch and K. C. Holmes, Proc. Natl. Acad. Sci. USA **88**, 5041 (1991).

164. L. Holm and C. Sander, Nature **361**, 309 (1993).

165. C. Chothia, Nature **357**, 543 (1992).

166. M. Oobatake and T. Ooi, Prog. Biophys. Mol. Biol. **59**, 237 (1993).

167. D. Sitkoff, K. A. Sharp and B. Honig, J. Phys. Chem. **96**, 1978 (1994).

168. D. Chandler and H. C. Anderson, J. Chem. Phys. **57**, 1930 (1972); F. Hirata and P. J. Rosky, Chem. Phys. Lett. **83**, 329 (1981).

169. Y. Okamoto, Recent Res. Develop. in Pure Applied Chem. **2**, 1 (1998).

170. M. Gō, N. Saitô and M. Ochiai, J. Phys. Soc. Jpn. **22**, 227 (1967); *ibid* **28**, 467 (1970).

171. E. J. Milner-White and R. Poet, Trends Biochem. Sci. **12**, 189 (1987).

Index

α-helix, 1, 6, 7, 9–11, 15, 18, 40, 47, 75, 83, 100, 107, 109, 113

β-sheet, 1, 113

β-strand, 1, 7, 8, 10, 11, 16, 47, 83, 117, 131

  standard ——, 17

β-structure, 6, 7, 40

  antiparallel ——, 1, 7, 8, 17, 27, 37, 72, 75, 95, 131, 133

  parallel ——, 1, 7, 8, 85, 87

Anfinsen's dogma, 21, 24, 97, 99, 104, 105, 109–111

Bremermann's method, 50, 97, 115, 125

CD, 29, 34

  stopped flow ——, 29

chaperonin, 110

comparative method of prediction, 113

cooperativity, 21, 34, 118

denaturation-renatureation phenomena, 32

distance map, 36, 40, 44, 95, 108

disulfide bond, 48, 62, 63, 75, 89, 97

DME, 36, 61, 98

DSSP program, 8, 90

evolution, 4, 108, 114

exon, 4, 108

fold recognition method, 114

folding pathway, 23

gene, 4, 108

genetic information, 1

  first deciphering of ——, 3

  second deciphering of ——, 5, 114

genome, 4, 5

globular protein, 4, 111

helix-coil transition, 6, 8, 21, 113, 117, 119

hemoglobin, 1, 4, 108

homology, 15, 17, 58, 114

homology recognition method, 114

hydrogen bond, 7, 8, 10, 18, 19, 24, 87, 97, 109, 117, 118

hydrophobic core, 24, 28, 32

hydrophobic interaction, 20, 26, 43, 45, 50, 95, 97, 99, 100, 105, 109, 110, 115

intron, 4, 108

island model, 10, 37, 47, 76, 90, 112

keratin, 4, 6

143

lampshade, 64, 73, 92, 97
Levinthal paradox, 23, 24, 28, 97, 121
long-distance interaction, 21
long-range interaction, 28
lysozyme, 5, 29, 33, 35, 40, 44, 62, 65, 69, 131

membrane protein, 4, 111
method of prediction, 5
    ab initio ——, 5, 113
    comparative ——, 113
    fold recognition ——, 114
    homology ——, 113
    threading ——, 114
module, 36, 108
molecular chaperone, 109, 110
molten globule state, 23, 27, 29, 32, 42, 45, 110, 118, 131
myoglobin, 4, 48, 50, 58, 59, 114, 115

N-D transition, 22
NMR, 5, 29, 32, 107, 115

packing order, 96, 98
phase transition, 9, 37, 107
    diffuse ——, 21, 37, 107
polymerase chain reaction, 4
prediction, 14
    —— of α-helix, 14, 113
    —— of β-bend, 113
    —— of β-strand, 14
    —— of secondary structure, 10, 15, 17, 113, 115

—— of tertiary structure, 5, 113, 114
    ab initio method of ——, 95, 97, 113
primary structure, 1, 100, 111, 112, 114
protein folding, 20, 27, 28, 33, 37, 110, 114
    chaperonin-mediated ——, 110
    first principle of ——, 21, 104
    mechanism of ——, 5, 18, 47, 95
    second principle of ——, 24, 28, 104

quaternary structure, 1, 5

random coil state, 4, 7, 8, 11, 32, 34
reversible denaturation and renaturation, 20
RMS, 36

secondary structure, 1, 6, 29, 47, 89, 90, 95, 110, 113
    standard ——, 8, 17, 61, 92
sequence homology, 114
simulated annealing, 50, 102, 115

tertiary structure, 1, 4, 5, 29, 104, 112–114
threading method, 114
two-state description, 29, 32
two-state theory, 32, 42, 45

X-ray, 4, 5, 8, 29, 32, 60, 89, 110